Soil Mechanics in Foundation Engineering

Volume 1

Soil Mechanics in Foundation Engineering

Volume 1 Properties of Soils and Site Investigations

ZENON WIŁUN
Professor of Foundation Engineering
Technical University of Warsaw
and
KRZYSZTOF STARZEWSKI
Ph.D., B.Sc., C.Eng., M.I.C.E.
Lecturer in Civil Engineering
The University of Aston in Birmingham

A Halsted Press Book

John Wiley & Sons
New York—Toronto

Published in the U.S.A. and Canada by Halsted Press, a Division of
John Wiley & Sons, Inc., New York

First published 1972

ISBN 0 470-82076–4

Library of Congress
Catalog Card No. 72–5120

Printed in Great Britain

Preface

Our purpose in writing this book has been to present the fundamentals of soil mechanics and foundation engineering in a manner understandable to all those concerned with the design and construction of foundations. Only a small group of men specialize in this field, yet, because foundations form an essential part of every building, structure or pavement, it is important to students and professional men in the fields of civil, structural, and municipal engineering, building engineering, and architecture. The common need of all is an understanding of the fundamental principles and problems encountered in the field as a whole. For those directly concerned with the design and construction of foundations it is essential to be able to apply the fundamental principles and design methods to an inexhaustible variety of practical problems. Thus, this book is an effort to make available a unified presentation of the fundamental principles of soil mechanics (Volume 1) and their application to real foundation problems (Volume 2).

All foundation problems, such as the determination of allowable bearing stresses or analysis of stability of slopes, are very complex and require a comprehensive knowledge of the material contained in both volumes of this book. The determination of, for example, allowable bearing stresses involves the following:

(1) Investigation of soil and ground water conditions on the site.
(2) Geotechnical analysis of physical and mechanical properties of the soils.
(3) Analysis of the problem on the basis of the findings from the above investigations and with reference to the proposed structure.

In Volume 1 the contents of Chapters 1, 3, and 6 cover the problems associated with site investigation work, identification of soils, and geological interpretation of the results; Chapter 4 covers the problems associated with the groundwater. Classification of soils and the determination of their physical properties is dealt with in Chapters 3 and 7; Chapter 2 deals with special problems associated with the mineralogical composition of soils and discusses interparticle forces in cohesive soils. Determination of mechanical properties of soils is discussed in detail in Chapters 5 and 6. A graphical correlation method for the determination of generalized characteristic soil properties is presented in Chapter 6.

In Volume 2 methods of evaluation of overburden stresses in soils and the determination of stress increments induced by applied loading are presented in Chapter 1. Methods of determination of allowable bearing stresses, with the consideration of ultimate bearing capacity and settlement and rates of settlement, are given in Chapter 2 and are further illustrated in Appendix A. Stability of natural and artificial slopes is discussed in Chapter 3 and determination of thrust on retaining structures and their stability in Chapter 4. A discussion of the effects of frost and preventive measures is contained in Chapter 5. Chapter 6 deals with the very important problem of compaction of soils. Chapter 7 contains many useful suggestions for design engineers and contractors and for observations of the settlement of structures. A comprehensive list of references, some of which are probably quite new to the western reader, are included in both volumes.

Our thanks are due to Mr. John Corbett and Mrs. Audrey Bennett, who have kindly read the script and to Mr. D. H. Bennett for his many valuable comments. We are indebted to Dr. R. S. Johnson of the Department of Geology at the University of Aston in Birmingham for his valuable comments on the contents of Chapter 1. We are also indebted to many members of the Civil Engineering Department at the University of Aston who have directly or indirectly assisted in the preparation of this book. We are particularly grateful to Mrs. Helena Starzewska who, apart from being a constant source of encouragement, has typed the script.

Extracts from British Standard Code of Practice CP2001 (1957), *Site Investigations*, British Standard Code of Practice CP2004 (1971, Draft), *Foundations*, and British Standard 1377 (1967), *Methods of Testing of Soils for Civil Engineering Purposes*, are reproduced by kind permission of the British Standards Institution, 2 Park Street, London W1A 2BS.

Z. Wiłun
K. Starzewski
1972

List of Contents

List of Figures

List of Tables

1

Origin of Soils

1.1. Engineering Soils

In civil engineering the term soil is applied to the natural products of weathering and mechanical disintegration of the rocks which form the crust of the earth. Soil is of interest to civil engineers because it occurs within the zone of influence of the stresses induced by newly erected buildings and structures and may also be employed in the construction of engineering structures.

1.2. Weathering and Disintegration of Rocks

The present surface of the earth consists of rocks of various ages and these geological formations are usually covered with soils which are the product of the natural weathering of rocks. The process of weathering of the earth surface has been going on for approximately 4500 million years. During this long period of time soils have been continuously formed from rocks and then transported by rivers, wind, and ice and eventually deposited in the oceans where they are transformed into rocks again. This process is going on at the present day.

Soils form through physical, chemical, and organic weathering of the rocks and the mechanical disintegration of rock fragments during transport in streams, rivers, ice, and wind.

1.2.1. PHYSICAL WEATHERING

Most rocks are formed under conditions very different from those pertaining to the earth's surface. Some are formed at high pressures, others at both high pressures and temperatures, while lavas are formed by cooling on the surface. All these processes induce stresses within the rock masses which may subsequently dissipate but, in general, will remain locked in.

When erosion removes great thickness of overlying rocks, the vertical overburden stresses on a particular rock body are gradually reduced and elastic expansion takes place. The rock mass is usually free to expand in the vertical direction only and therefore the horizontal overburden stresses dissipate at a much lower rate than the vertical stresses. This differential rate of stress relief leads to a build-up of high horizontal stresses near the ground surface and to the development of cracks (also called joints) parallel to the surface. As the result, exfoliation and spalling takes place of the rock near the surface. At the same time, any relief in the horizontal direction develops vertical joints.

If water penetrates any fissure in rocks, it may freeze in the higher latitudes. Water increases its volume on freezing and the resultant wedging action leads to a further splitting and disintegration of rock.

1.2.2. CHEMICAL WEATHERING

The physical processes of fragmentation of rocks and of crack formation result in a considerable increase in the surface area of the rocks exposed to the air and in a deeper penetration and accumulation of the rain water. These factors help to accelerate the processes of the *chemical weathering* of the rocks.

Oxygen and carbon dioxide from the atmosphere, and organic acids from soil, dissolve in water and react with the rock to form at various rates new compounds such as clay colloids, silica, carbonates, and iron oxides. The decomposition of feldspar and mica leads to formation of clay particles (clay colloids), whereas highly resistant quartz grains remain unaltered and are washed out to form sands. Particularly susceptible to decomposition by the combined action of the rain water and dissolved oxygen from the air are the dark coloured rock minerals, usually rich in iron.

1.2.3. ORGANIC WEATHERING

Organic weathering is induced by the processes of the animal and plant life. A prominent role is played by bacteria which induce chemical changes in their surroundings. The number of bacteria is greatest on the ground surface, but at a depth of 1 m there is about 150 000 bacteria per 1 ml of soil and they only cease to exist at depths between 3 to 5 m. Some bacteria produce carbonic acid, others nitric or nitrous acid, ammonium, hydrogen sulphide, or marsh gas, all of which contribute to further weathering of the soils.

1.2.4. MECHANICAL DISINTEGRATION

Fragments of rock loosened by *in situ* weathering roll or are washed down to lower ground and are carried away by streams and rivers. During this transportation the fragments are further broken down and rounded by continuous abrasion and attrition.

Sea waves break against the cliffs like giant hammers, shattering the rock, penetrating joints and washing any loose fragments away; the rock fragments (rock debris) carried by the water are then further reduced in size and rounded by continuous grinding down as a result of wave action.

Movement of glaciers down mountain slopes, and, in particular, ice sheets over continental masses in higher latitudes, also contributes greatly to the general process of mechanical disintegration as the slow-moving ice, armed with embedded fragments of rock, abrades the bed-rock and grinds the fragments down. Plucking of rock masses by the movement of ice has an even greater effect.

1.3. Classification of Soils with Relation to their Origin

Depending on their origin the superficial deposits can be divided into residual soils and transported (sedimentary) soils: alluvial, marine, glacial, Aeolian, lacustrine, and organic.

1.3.1. RESIDUAL SOILS

The residual soils are formed *in situ* by weathering of the original rocks; they usually consist of stoney clayey sandy silts or, if the finer particles have been leached out, of residual rock debris.

Figure 1.1. Profile of residual soil: 1, top soil; 2, stoney clayey sandy silt; 3, clayey rock debris.

Figure 1.2. Differential erosion (by courtesy of Aerofilms Ltd).

In partially weathered rock, stoney clayey sandy silts are formed which consist of the unsoluble clay particles, a certain proportion of unaltered crystals and fragments, and carbonates. It is characteristic of this type of soil that the proportion of the rock fragments increases with depth until finally the natural unaltered rock is reached (Figure 1.1); the fragments are distinctly angular.

During erosion of the weathered rock, the fine clay and silt-size particles, with carbonates in solution, are removed and carried away by streams and rivers to the sedimentary basins; the remaining fragments of the original rock form 'residual rock debris'.

If the decomposition or disintegration of rock masses proceeds at different rates, then it is referred to as 'differential weathering'; areas of greater resistance project as ridges above the faster weathering zones (Figure 1.2).

In tropical countries residual soils known as laterites and bauxites are formed. The process of formation of these soils is greatly assisted by their alternating saturation and desiccation which helps to induce concentration of iron oxides and alumina in the upper horizons; lateritic and bauxitic clays are usually red in colour, soft when naturally moist, but harden on exposure to atmosphere.

1.3.2. TRANSPORTED SOILS

1.3.2.1. Alluvial Soils (Deposits Laid Down by Rivers)

The fast-flowing waters of mountainous streams are capable of transporting not only small particles but also large rock fragments. Owing to abrasion and attrition these fragments become ground down, the larger becoming rounded boulders, cobbles, and gravel-size pebbles, and the smaller sand grains. The very fine product of this process forms the silt fraction in many soils.

The sand grains and the silt particles are carried by rivers over vast distances, but may be deposited, often temporarily, whenever the velocity of the water decreases (Figure 1.3). The coarsest particles, i.e. gravels and coarse sands, form

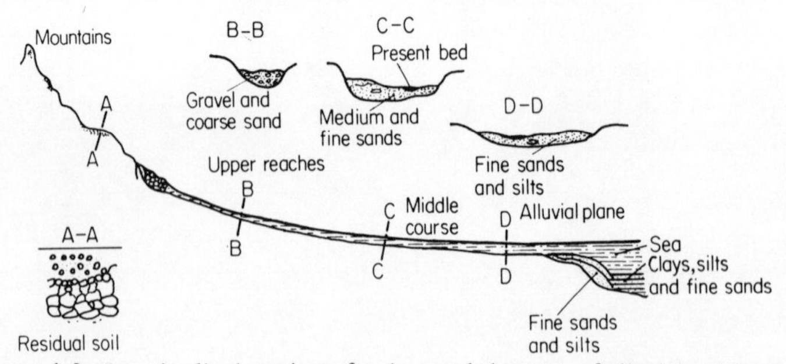

Figure 1.3. Longitudinal section of a river and the type of alluvial sediments.

deposits most commonly in the upper reaches of rivers, the medium sands in their middle courses, while the fine sands and silts are spread on the lower alluvial plains.

To appreciate the magnitude of the work done by the rivers in transportation of soil particles (sediment) one only has to look at the figure quoted for the River Danube: 900 000 m^3 of sediment per year.

The transported detrital load of rivers not only consists of the products of weathering of the rocks in their upper reaches, but also of material derived from the banks and river bottoms. The maximum depth of down-cutting of a river depends, to a large degree, on the level of the sea or lake into which it discharges. This is known as the base level. When the river has reached a natural gradient, a

Figure 1.4. Meandering river–note signs of previous positions of the river bed (by courtesy of Aerofilms Ltd.).

relative lowering of the base level will cause the river to cut down further owing to the newly acquired energy and this again lowers its bed. It is then said to be *rejuvenated.* As the river reaches its general base level, gradual silting of the river bed begins with the establishment of flood plains of alluvium (Figure 1.4).

1.3.2.2. Marine Deposits

The particles of clay and silt are carried by the rivers to the sea where they coagulate and are laid down to form deposits of mud. Interlayered with mud are beds of fine sand brought by the rivers during flood.

In the past, the British Isles were covered many times by seas in which the majority of the present solid rocks were deposited and since uplifted. For example, during the Eocene Period the London Clay was deposited in a fairly shallow basin which extended from Belgium and France to eastern England. Accumulations of shell fragments and calcareous algae in the shallow seas of the Cretaceous Period formed the well-known chalk deposits in the south-east lot England, while the Jurassic seas gave rise to the Lias clays, shales, and limestones of the present day.

Thus deposits of pure shell fragments and skeletons of marine micro-organisms gave rise to the chalk and other shelly limestones, while the presence of other detrital material lead to the formation of sandy and clayey limestones. Overburden pressures and changes in temperature and chemical environment transform sand deposit in these sea basins into various sandstones, whereas muds become mudstones and shales. Under the action of heat and high stress sedimentary rocks are metamorphosed to slate, schist, and gneiss. Subsequent earth movements may have uplifted these rocks so that they are now being weathered and eroded again to form new sediments.

1.3.2.3. Glacial Deposits

During the last million years, the British Isles have been subjected to alternating very cold and mild climates. During the most recent cold period, ice gradually spread out from the Lake District and Welsh and Scottish mountains, until sheets of ice covered most of Britain north of a line from the Bristol Channel to the

Figure 1.5. Valley glacier (by courtesy of Aerofilms Ltd.).

Thames Estuary. The climatic conditions in fact oscillated during this time and the ice sheet advanced and retreated several times. The thickness of the ice sheet at its maximum was in the region of 500 to 1000 m, which imposed stresses on the soils beneath it of up to 10 MN/m^2 (100 tonf/ft^2, 100 kgf/cm^2), i.e. much greater than the stresses imposed by the present-day structures.

The landscape in many parts of Britain was profoundly modified by the plucking of large masses of rocks which, embedded in ice, acted as a gigantic flexible file which removed and ground down soil and bed-rock. Eventually this material was deposited in front of the glacier as various moraine deposits.

When the ice sheet retreated during warmer periods, masses of the 'glacial drift' were deposited on the land surface in the form of boulders, often set in

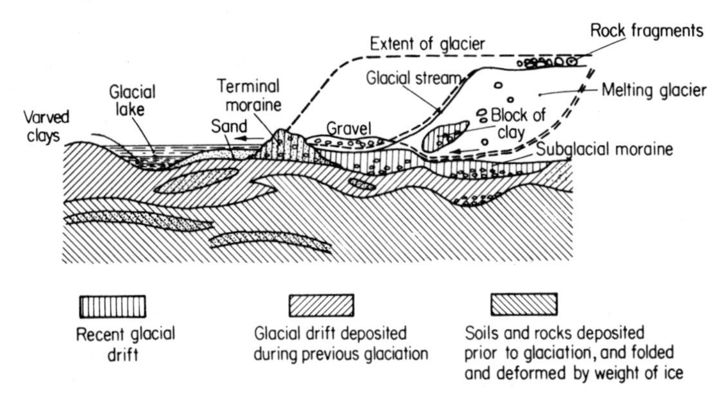

Figure 1.6. Longitudinal section through a retreating glacier.

a matrix of clay, with fluvio-glacial sands and gravels (Figure 1.6). Varved clays were formed in lakes in the vicinity of melting glaciers.

Boulder clays are unsorted deposits which usually consist of stiff clays composed partly of crumbled soil and rock, and partly of rock flour produced by the ice as it moved. Boulders and rock fragments of all sizes and types are embedded. Where glaciers flowed over calcareous rocks, the deposits have a high calcium carbonate ($CaCO_3$) content; when over shales and mudstones, a predominantly clay content. Varved clays consist of thin, alternating dark and light layers, the former consisting of clay and the latter silt.

Most of the soils deposited prior to the last glaciation period, with subglacial deposits, were consolidated (densified) by the weight of the overlying ice and are now characterized by low porosity and compressibility. They can sustain large bearing stresses and form excellent foundation material. However, some of these soils were frozen during the subsequent glaciations. Not only were these not consolidated by the weight of the overlying ice, but they may have been dilated during the freezing process. Thus their porosity and compressibility may be even higher than it would have been in their normally consolidated state. In other localities soils deposited during the retreat of the ice sheet have remained in the normally consolidated state and, because of their variable nature, are not good foundation material.

In other areas, soils consolidated during one glaciation have been considerably deformed and folded, not only by the weight of the advancing and retreating glaciers of the subsequent glaciations, but also by various periglacial phases. Solifluction deposits and associated slip planes have often developed in cohesive soils. The shearing resistance of such soils is considerably reduced by the presence of these slip planes (see Volume 2, Chapter 3).

1.3.2.4. Aeolian Deposits (Wind-blown Deposits)

As the ice retreated, vast areas of barren land were exposed. Large quantities of silt particles (dust) were picked up from this surface by strong winds and transported long distances. When the intensity of the winds subsided, these dust fractions settled to form loess deposits. Often these wind deposits are cemented (e.g. with calcium carbonate) immediately subsequent to their deposition. Their porosity is normally fairly high (46 to 52%).

In a dry state, loess soils have a reasonably hard consistency and possess considerable strength. Deep cuttings with almost vertical faces confirm this. However, loess is easily eroded and streams and rivers cut deep gorges. When loess becomes saturated beneath a foundation, sudden additional settlement occurs as the cementing agent holding the grains in position is dissolved and the loose structure collapses. Loess soils under desert conditions have exhibited additional settlements of up to 1·5 m. On the other hand, loess deposited under water, or with the carbonate matrix already leached away, does not possess the properties of a metastable soil because it is normally consolidated.

Winds are also capable of moving dry sands to form sand dunes; mobile dunes which reach a moist ground gradually attenuate as the fine sands, wetted by the capillary rise of moisture, become immobile.

1.3.2.5. Organic and Stagnant-water Deposits

The ground surface continuously undergoes change under the eroding action of wind and rain water. At the same time sediments accumulate in inland reservoirs and lakes, on river flood plains, and in estuaries, where mineral particles are deposited to form lacustrine or alluvial muds. Very often considerable quantities of organic matter ($> 5\%$) are deposited at the same time to form organic muds, also known as organic clays or silts.

Some lakes (e.g. post-glacial) and old river beds (e.g. ox-bows) become overgrown and change into peat bogs. The thickness of peat and organic muds formed on the river flood plains is not great and does not generally exceed 4 to 5 m. Peat formed in post-glacial lakes can be 10 to 20 m thick.

1.3.2.6. Agricultural Soils

All immature soils are subject to the continuous process of weathering. Particularly pronounced changes have taken place, and are continuously taking place, in the layers close to the ground surface and lead to the formation of the agricultural soils. An agricultural soil forms the top 1 to 2 m and supports plant growth.

During *in situ* investigations of soils it can easily be observed that (in different and far-apart spaced trial pits) the upper layer of soil exhibits distinctive horizontal bands of similar colour and granular composition. The uppermost horizon is usually dark in colour and is referred to as the fertile soil. Beneath it is usually found a light, almost colourless soil frequently of silty sandy nature. Deeper lies a layer of soil of distinct and bright colouring. These three, or in

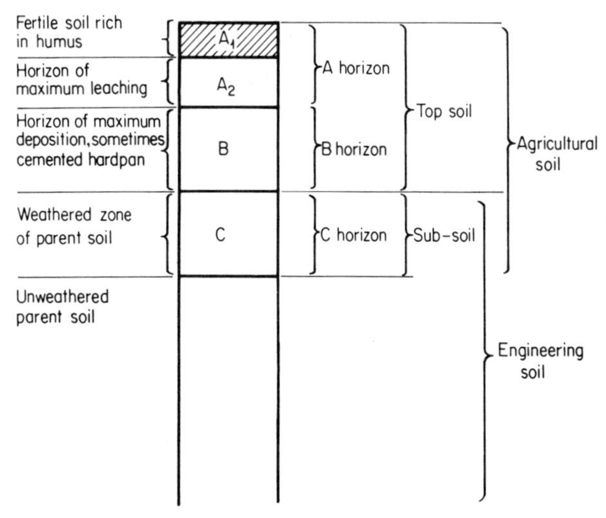

Figure 1.7. Relation between agricultural and engineering soil (based on British Standard Code of Practice CP 2001 (1951)).

some cases, further layers, have been derived from the same underlying parent soil and in engineering terms are referred to as the top-soil layers (Figure 1.7).

The process of formation of agricultural soil proceeds as follows. After deposition or formation, its surface becomes subject to bacterial activity. The soil undergoes further chemical change and gradually becomes capable of support of vegetation which in turn, on dying, enriches it in organic constituents. This leads to formation of the various organic compounds known as humus. As the humus content increases, the surface layer of the soil becomes more fertile and darker in colour; it forms the uppermost layer of the A (eluvial) horizon.

As formation of the fertile soil proceeds, rain water leaches out mineral salts and colloidal particles from the underlying soil. These colloidal particles and salts are brightly coloured but when they are leached away the soil becomes colourless to form the major layer of the A horizon.

The third layer, known as the layer of maximum deposition, retains most of the leached-out minerals. Its colloidal fraction content is increased and it often becomes brightly coloured. The soil in this layer, depending on the type of the minerals retained, becomes more clayey or more cemented than the underlying parent soil. It is known as the B (illuvial) horizon of the top soil.

The weathered zone of the parent soil underlying the top-soil is known as the C horizon or the sub-soil. In clays a characteristic columnar structure may be present.

Agricultural soils affected by waterlogging are called gley soils. Where the effects are confined to particular horizons within the profile these are called gley horizons. Waterlogging retards the processes of oxidation of iron compounds and leads to the development of the characteristic colour range of the gley soils, varying from dark greys to bluish greens. In site investigation work the recording of a gley horizon is of particular importance because it indicates the level to which the ground water rises. Apart from that, the bluish and greenish colours indicate the presence of the dangerous, heave-susceptible soil.

In low-lying, waterlogged areas, marshy or boggy soils are frequently encountered. They are characterized by the presence of a layer of peaty soil above the gley horizon.

2

Basic Physico-chemical Properties of Soils

The safe bearing capacity of a soil should be determined from a knowledge of its physical (Chapter 3) and mechanical (Chapter 5) properties.

The physical and mechanical properties of soils, particularly in the case of clays, depend to a large extent on their physico-chemical characteristics, of which the following are of particular importance.

(a) Mineralogical composition.
(b) Nature of the soil particle surface.
(c) Physico-chemical activities at the solid–fluid interface.
(d) Capillary effects.

2.1. Mineralogical Composition of Soils

From the content of the previous chapter it can be seen that granulometric compositions of soils may contain (a) boulders and cobbles, (b) gravel-size rock fragments or pebbles, (c) sand grains, (d) silt particles, (e) clay particles, and (f) particles of organic matter in different stage of decay.

Boulders, cobbles, and *gravel-size particles* have the same mineralogical composition as the parent rock (Table 2.1) from which they were derived by

Table 2.1. Average mineralogical composition of igneous and sedimentary rocks (after Kirsch, 1968)

Igneous rocks		Sedimentary rocks	
feldspar	59%	quartz	30%
augite and hornblende	17%	mica	23%
		clay minerals	17·5%
quartz	12%	feldspar	9%
mica	4%	carbonates	8·5%
others	8%	Fe_3O_3, etc.	5·5%
		chlorite	2%
		water	2%
		others	2·5%

weathering processes or by mechanical disintegration. Some boulders and most cobbles and pebbles have a weathered surface layer which can usually be recognized by lighter colouring and lower strength than the unweathered inner core.

In some cases, when weathering has penetrated the inner core of a rock fragment, its strength is so small that a relatively light pressure will break it up into small fragments consisting of materials more resistant to chemical weathering.

Sand grains are generally composed of materials resistant to weathering such as quartz and silica. In recently deposited sands, grains of feldspar and mica which are much less resistant to weathering can be found. In some parts of the world mica, gypsum, and limestone sands are known to exist. Their grain crushing strength is small and they exhibit high compressibility.

Silt particles (rock flour) are the product of grinding action between larger rock fragments during their transportation by ice, water, or wind. Recent silts contain, in addition to quartz and silica, particles of feldspar and mica. However, the latter two are subject to rapid chemical weathering and are either washed away or recrystallized as clay minerals and remain *in situ* to form clayey silts.

Clay particles (<0·002 mm) mainly consist of clay minerals which are the product of chemical weathering of feldspars and micas. In the process of chemical weathering of feldspars, the cations go into solution and owing to hydrolysis or

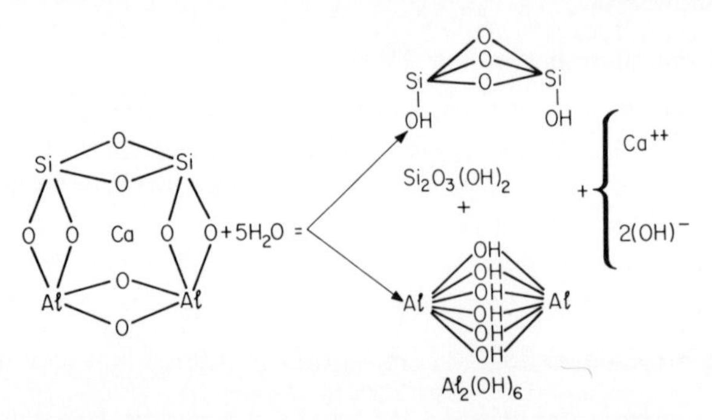

Figure 2.1. Example of chemical weathering of anorthite.

hydration there is a partial decomposition of the aluminosilicate anions. Decomposition of an anorthite (Figure 2.1) can be taken as an example of a complete decomposition of feldspar.

On removal of calcium the newly formed aluminium hydroxides and silicates combine through crystallization and form clay minerals (Figure 2.2) with the characteristic layered crystalline structure.

The basic crystal layers of clay minerals consist of tetrahedral silica sheets (silicon dioxide) arranged in a hexagonal basal pattern. These are combined with

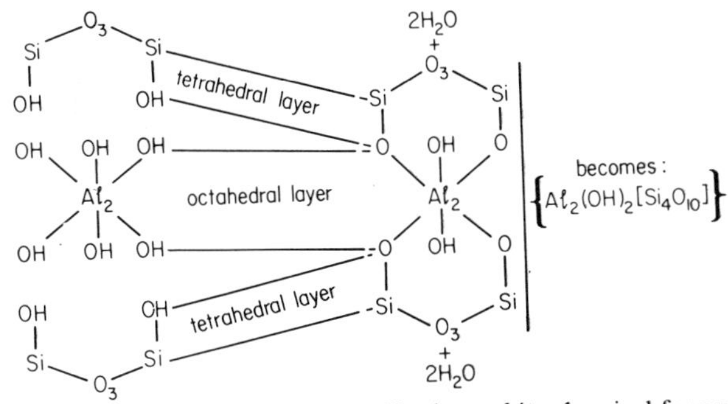

Figure 2.2. Formation of montmorillonite and its chemical formula.

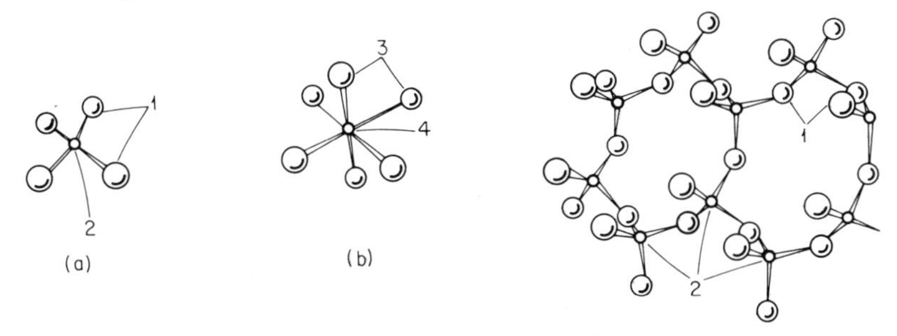

Figure 2.3. Schematic arrangement of structural elements of clay mineral crystals; (a) tetrahedron SiO_4; (b) octahedron $Al(OH)_6$; (c) arrangement of tetrahedrons into hexagons; 1, oxygen atoms (large spheres); 2, silicon atoms (small spheres); 3, hydroxide molecule (large spheres); 4, aluminium atom (small sphere).

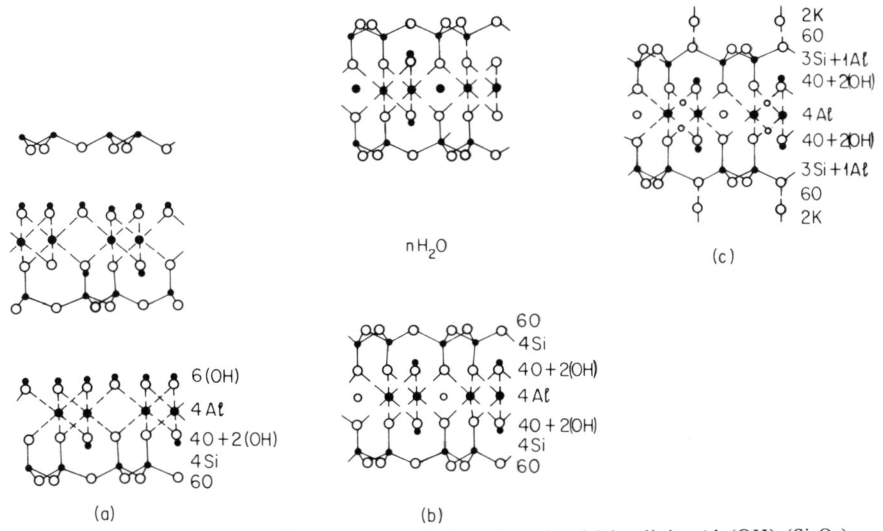

Figure 2.4. Atomic crystalline structure of clay minerals: (a) kaolinite $Al_2(OH)_4(Si_2O_5)$; (b) montmorillonite $Al_2(OH)_4(Si_4O_{10}) + nH_2O$; (c) muscovite $Al_2(OH)_2(Si_3Al_{10})$.

either one or two octahedral aluminum hydroxide sheets—also known as gibbsite sheets (Figure 2.3).

Depending on the arrangement of the basic layers, and on the presence of any additional elements, different clay minerals are formed (Figure 2.4). Clay minerals can be divided into three main groups.

(a) Kaolinite.
(b) Montmorillonite.
(c) Illite.

Kaolinite minerals are the main constituents of china clays and kaolin. Their basic layers consist of a silica sheet combined with a gibbsite sheet which are

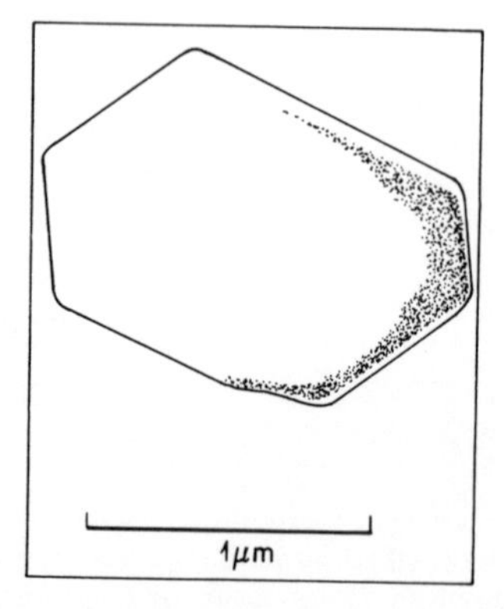

Figure 2.5. Kaolinite particle (after Lambé, 1951).

strongly bonded together to form a compact crystalline lattice. Kaolinite particles are flat approximately 20 nm thick and between 100 and 500 nm wide (Figure 2.5). Water absorption of kaolinite minerals at saturation is in the region of 90% of dry weight.

Montmorillonite minerals are mainly to be found in bentonites and fullers' earth. The basic layers consist of a gibbsite-structured sheet sandwiched between two tetrahedral silica sheets (Figure 2.4). With such arrangement of molecules, the outside surfaces consist of oxygen atoms only, resulting in very weak bonds between the basic layers. On hydration these basal cleavage planes admit water molecules into the crystalline lattice and cause swelling; the amount of water absorbed into the crystalline structure in this way can be high enough to more than double the particle thickness.

Montmorillonite particles are approximately 1 nm thick and 200 nm wide. Water absorption of these particles at saturation is between 300 and 700%.

The crystalline structure of illites is basically similar to that of montmorillonites (Figure 2.4) but potassium ions are present between the basic layers, compensating their out-of-balance electrostatic charge and strengthening the bonds between them.

Water absorption of illite particles is half-way between that of kaolinites and montmorillonites. Dimensions of illite particles are as follows: thickness, around 10 nm; width, around 200 nm.

The clay fraction may also contain, apart from clay minerals, very small particles (<0·002 mm) of chemically unaltered quartz, feldspar, or mica. The proportion of these particles in any soil will depend on the conditions under which it was formed.

In northern regions, where the mean annual temperature is below +5 °C, only part of the feldspars and micas will be decomposed by chemical weathering; however, in tropical regions the weathering will affect all feldspars and micas.

During different geological periods different climates existed in the British Isles, ranging between extreme polar and tropical conditions; hence, the soils encountered contain different proportions of the unweathered clay-forming minerals.

Organic particles are derived from decomposition of the remains of plants and animals, also from bacteria and fungi-assisted chemical reactions between the decomposition products and the soil's mineral compounds. The newly formed particles have an amorphous structure, and their water absorption is in the order of 500% of their dry weight.

2.2. Surface of Soil Particles

Soils consist of individual grains and particles which form the soil skeleton. The voids between the grains or particles are filled with water or partly with water and partly with air (water vapour or gases). In any case water covers the surface of the soil particles and the air is present in the form of larger or smaller bubbles contained wholly in the water.

The surface of the soil particles, i.e. the interface between the solid phase and the liquid phase (water or solution of various chemical compounds), is the seat of many physico-chemical phenomena (adsorption of water and ions, electrokinetic potential, ion exchange capacity, cohesion, etc.). These phenomena affect the nature and behaviour of a given soil; they dictate its structure, compressibility, and strength. Any possible strengthening of a given soil, by either physical or chemical means (stabilization of soils), is also governed by these phenomena.

The intensity of physico-chemical phenomena varies for different soils and depends on the mineralogical composition of the soil particles, on chemical composition of the pore water, and also on the surface area of the particles.

The total surface area of all the soil grains or particles divided by their volume or mass is known as the *specific surface.*

The smaller the particles of a given soil the larger is its specific surface (Figure 2.6) and hence the greater is its physico-chemical activity. A sand consisting of grains of approximately 1 mm in diameter will have a specific

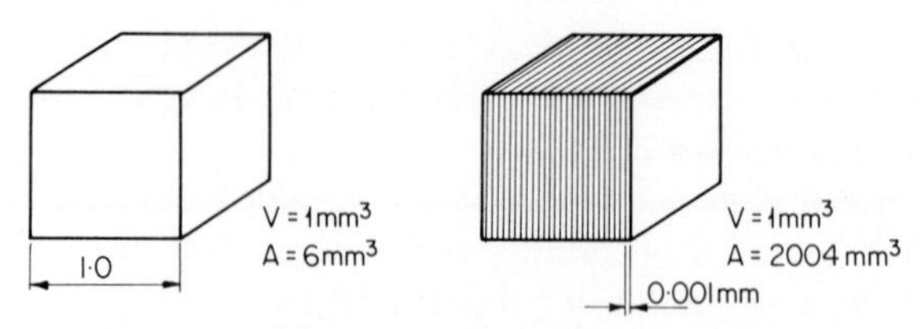

Figure 2.6. Surface area of a cube and of a set of plates of the same volume.

surface around 6 mm^2/mm^3, while the specific surface of clays containing more than 30% of clay fraction (particles less than 0·002 mm in size) will be many thousand of times greater. If the specific surface is evaluated per unit mass, then for the above sand it can be taken as 2000 mm^2/g (0·002 m^2/g).

Typical specific surfaces of clays according to Priklonskij (1949) are given in Table 2.2.

Table 2.2. Specific surfaces of clays

Type of soil	Specific surface (m^2/g)
kaolin	80
glauconite	400
black earth	440–990
bentonite	1300–1390

2.3. Physico-chemical Phenomena at the Interface

2.3.1. ACTIVITY OF THE SOLID–FLUID INTERFACE

Soil particles are built up of ions of different elements. The ions within any given particle (inside the crystalline lattice) are completely balanced, their bonds being satisfied by the surrounding ions.

On the other hand, the ions located within the surface are only balanced on the inside, whilst on the outside they are electrically unbalanced and tend to combine with other ions or molecules which happen to be within the range of their molecular attractive forces. Consequently, soil particles are capable of attracting and holding water molecules and ions on their surface, ions of opposite sign to those present in the surface of the crystalline lattice. This leads to formation of an adsorbed water layer and a double ionic layer around the particle.

The adsorbed water is very strongly attracted to the surface of the soil particles; the water molecules immediately in contact with the surface are believed to be attracted to it with a force of the order of 2500 N/mm^2 (355 000 lb/in^2, 25 000 kgf/cm^2) which gives them characteristics of a solid body of density 1·7 g/ml (1·7 g/cm^3) (Rode, 1955). The force with which the water molecules are attracted to the surface of a particle diminishes with distance and according to Winterkorn (1955) is of the order of 5 N/mm^2

(710 lbf/in^2, 50 kg/cm^2) for saturation equivalent to the hygroscopic moisture; for complete saturation this force diminishes to zero (Figure 2.7).

Within the adsorbed water layer, water molecules (as diapoles*) are orientated in such a manner that the positive poles are facing the anions on the surface of

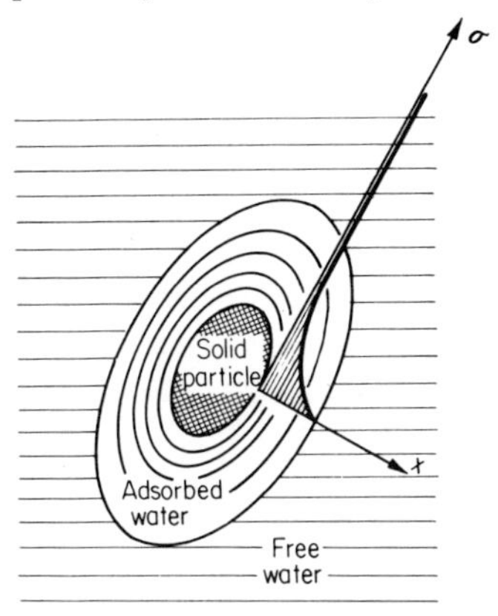

Figure 2.7. Distribution of attractive forces within the adsorbed water layer: σ, attractive forces; x, distance from the particle surface.

the particles, and the negative poles are facing the cations present in the adsorbed water layer or positive poles of the adjoining water molecules.

Double ionic layer. Because a solid particle is surrounded by the oriented water diapoles, the electric charge of the surface ions is not electrostatically balanced (diapoles in themselves are electrostatically neutral). To achieve the balance, ions must be adsorbed from the water solution present in the pores.

The ions present in the surface of the crystalline lattice of a solid particle may initially possess charges of opposite signs; in the first instance, therefore, there will be an equalization of the sign of the surface charge either through solution of ions of one type from the lattice or through adsorption of appropriate ions from the solution. As the result of either of the changes (depending on the solubility of the solid phase) equalization of the sign of the surface charge will be achieved. In moderate climatic conditions anions will usually be found within the surface of soil particles while in laterites, formed under tropical conditions, the surface ions are predominantly cations.

In order to balance electrostatically the anions present within the surface of particles, cations are adsorbed from the water solution in the pores. However, the attracted cations cannot form a single layer around a particle, because they themselves are surrounded by water molecules which prevent them from packing closely together, and hence they are distributed (diffused) through a certain layer, with the greatest concentration close to the particle surface, gradually decreasing until the normal concentration of the pore water solution is reached.

* Diapole is the arrangement of two equal electrostatic charges of opposite sign.

This layer is known as a diffuse layer (Figure 2.8). As the concentration of cations decreases the anions become more numerous, reaching the normal concentration together with cations. The diffuse layer of cations together with the surface layer of anions is called a double diffuse layer.

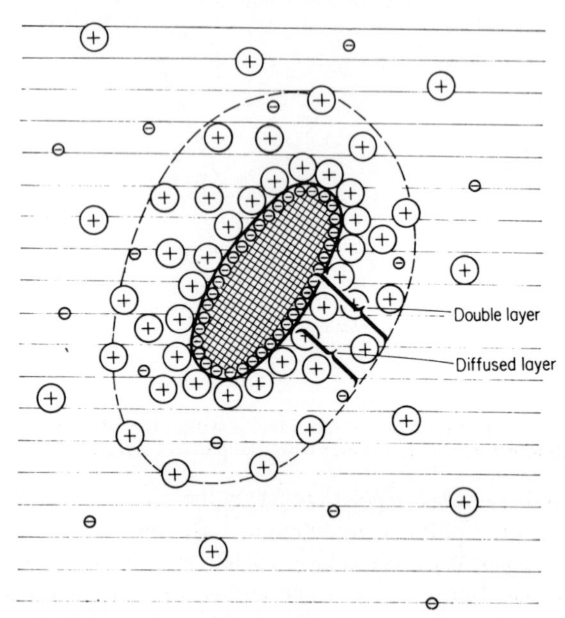

Figure 2.8. Particle surrounded by a double diffuse layer.

The thickness of the double layer, and of the adsorbed water layer, depends on the chemical composition of the solid particles and on the valency of the adsorbed ions. The type of adsorbed cations has a very significant influence on the behaviour of the soil; the greater the valency of the adsorbed cations, the better are the mechanical properties of a given soil.

Clays containing sodium cations are characterized by high water absorption and considerable swelling; after replacement with cations of greater valency, e.g. calcium, the thickness of the double layer decreases, the particles coagulate and the clay exhibits considerably reduced swelling. As an example bentonite with adsorbed sodium cations (Na-bentonite) has a water absorption in the region of 700%, while the same bentonite with calcium cations (Ca-bentonite) has water absorption of only 300%.

One of the methods of strengthening of clays can be the exchange of cations; the higher the valency of the introduced cations the better are the results (Figure 2.9). The exchange of cations can often be carried out electrochemically by passing a direct current through the soil between aluminium anodes and copper cathodes. Bernatzik (1947) states that with the above method he has

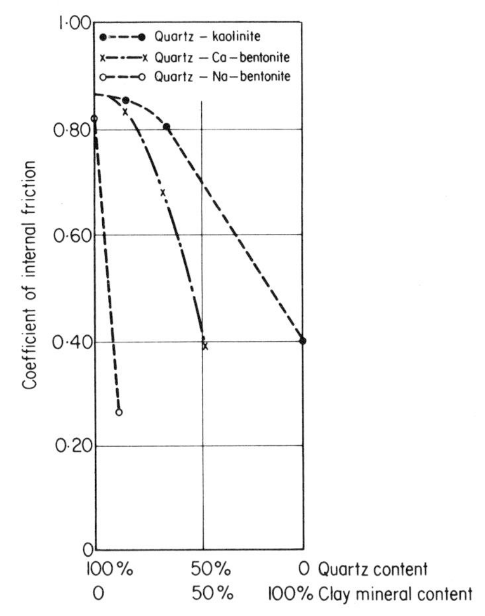

Figure 2.9. Relationship between the coefficient of internal friction of a soil and its mineralogical composition and the type of adsorbed cations (Endell *et al.*, 1939).

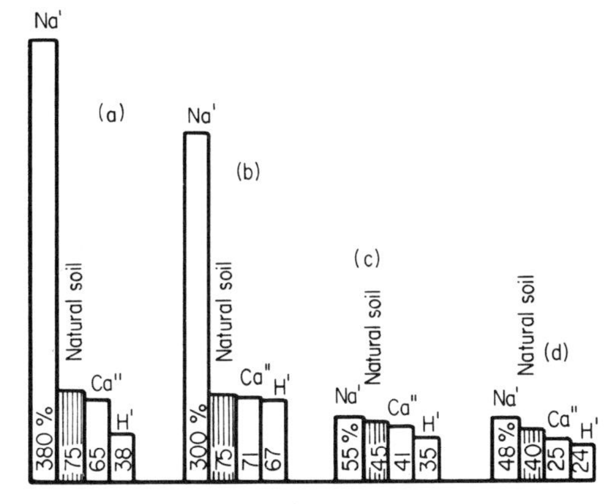

Figure 2.10. Influence of adsorbed cation on the swelling of different soils on saturation with water, expressed as percentage of the original volume (after J. S. Kulish (Babkov, 1956)): (a) black earth; (b) alluvial silt; (c) loess; (d) boulder clay.

observed increases in the angle of internal friction from 23° to 35° with a simultaneous substantial decrease in the compressibility of the soil.

The influence of adsorbed cations on the swelling of different soils is shown in Figure 2.10 (Babkov and Gerburt-Gejbovich, 1956). The type of cation and the thickness of the adsorbed water have a considerable effect on the permeability of soils; the thicker the absorbed water layer surrounding the surface of soil particles, the smaller is its permeability because a greater proportion of the volume of the pores is occupied by the very strongly held adsorbed water.

2.3.2. ION EXCHANGE CAPACITY

Soils with highly developed surface activities are capable of the adsorption and retention of ions of minerals dissolved in the water, while at the same time they release into solution an equivalent quantity of ions contained in their double layer. The quantity of the exchangeable ions expressed in milliequivalents per 100 g mass of the dry soil is known as the ion exchange capacity.

For natural soils the ion exchange capacity varies between 0 and 40 milliequivalents per 100 g; for clay minerals the limits are as follows:

for kaolinite	3–15	milliequivalents/100 g
for illite	20–40	milliequivalents/100 g
for montmorillonite	60–100	milliequivalents/100 g

There is a definite relationship between the ion exchange capacity and the plasticity index of cohesive soils. According to Piaskowski (1954) this relationship can be taken as linear (Figure 2.11).

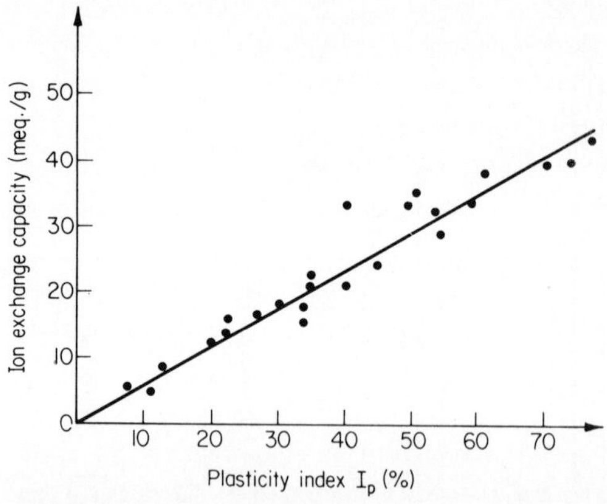

Figure 2.11. Relationship between plasticity index and ion exchange capacity.

2.3.3. ELECTROKINETIC POTENTIAL

In connection with the variation of concentration of cations in the diffuse layer there is a certain drop of potential across it. The potential drop between the innermost layer of the adsorbed cations (adjacent to the particle surface) and the outermost layer of the cations in the diffuse zone is known as the electrokinetic potential (Figure 2.12). The existence of this potential results in many electrical and electrokinetic phenomena which have an important bearing on the behaviour of soils.

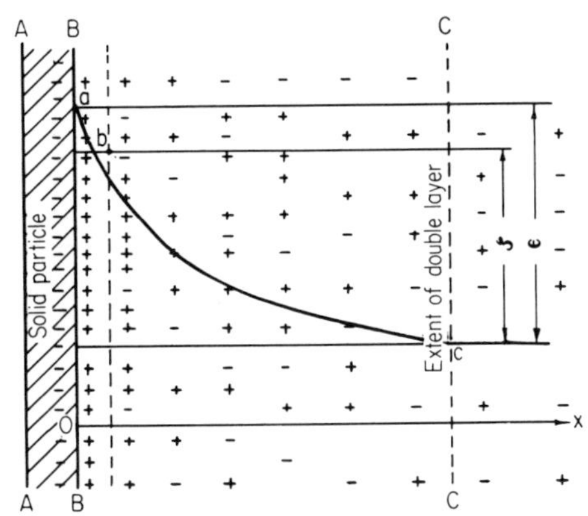

Figure 2.12. Double layer: AC, double layer; BC, diffuse layer; abc, potential drop in the double layer; bc, potential drop in the diffuse layer, i.e. the electrokinetic potential (Preece, 1947).

The magnitude and sign of the electrokinetic potential depend on the mineralogical composition of the soil particles, on the water content of the soil, on the type and quantity of ions present in the pore water, and also on the temperature of the soil.

Existence of the electrokinetic potential can easily be verified by passing a direct current between two electrodes placed in a beaker containing the clay suspension. Clay particles, having a negative electrokinetic potential, will behave as anions (during electrolysis) and will move with a certain velocity towards the anode. Their velocity will depend on the magnitude of the potential (WGT Gidroproject, 1950) as follows:

$$V = \zeta/21 \tag{2.1}$$

where V = electrophoretic velocity (μm/s per 0·1 V/mm)
ζ = electrokinetic potential (mV)

The sign of the electrokinetic potential depends on the pH value of the clay suspension and on the pH_{is}* value of the given soil. When $pH > pH_{is}$, the potential ζ is negative; when $pH = pH_{is}$, then $\zeta = 0$; and when $pH < pH_{is}$, then the potential ζ is positive. For most soils in the British Isles $pH_{is} < 7$, and therefore under normal conditions their potential will be negative. On changing the value of the pH by the addition, of, say, a few drops of hydrochloric acid to the suspension, it is possible to change the sign of the potential ζ to the positive.

If all the particles in a suspension have a potential of the same sign, then they will repel each other in the same manner as similarly electrostatically

* pH_{is} value of a given soil is defined as the pH value of a particular suspension of the soil in which, on application of an electric potential, there is no movement of particles towards any of the electrodes.

charged glass balls repel each other. The greater the electrokinetic potential the stronger is the repulsion between individual grains. On the addition of hydrochloric acid to a suspension its *p*H value changes, which should result in the change of the sign of the electrokinetic potential ζ; the change in the sign can only take place by going through a stage when $\zeta = 0$.

At the instant when $\zeta = 0$ (or is sufficiently small), the repulsive forces between the particles in the suspension are almost eliminated and the attractive forces take over, leading to coagulation of the particles, the neighbouring particles join into small groups (flocs) which precipitate to form a jelly-like sediment.

Hydrochloric acid and sodium chloride are typical electrolytes which produce coagulation by decreasing the potential ζ. When it is required to disperse soil particles, i.e. to break up existing flocs (e.g. in sedimentation grain size analysis), it is necessary to increase the potential ζ by addition of a suitable electrolyte, called a stabilizer. Common stabilizers are sodium oxalate, ammonium solution, or sodium hexametaphosphate (recommended in British Standard 1377 (1967)).

It must be remembered that coagulation or stabilization of a suspension is accompanied by many other phenomena such as exchange of ions, decrease in thickness of the adsorbed water layer and double diffuse layer, and change in temperature, etc.; these phenomena will affect the reaction. It is frequently necessary in stabilization to carry out several trials, using different additives and varying the quantity of these additives in order to prevent coagulation of suspension of a given soil.

2.3.4. ELECTROKINETIC PHENOMENA

The pores within a soil form a continuous network of channels of varying cross-sectional area. The walls of these channels, as in the case of individual particles, are lined with the double ionic and the adsorbed water layers.

Movement of water in the fine-grained soils through these channels is considerably obstructed by the presence in the pores of the strongly held and densified adsorbed water. If, however, direct electric current is passed through the soil, then owing to its action part of the ionic diffuse layer will be tangentially displaced from the rest of the adsorbed layer and will take with it all the water molecules attracted by the individual ions, towards the electrode of the opposite sign. This phenomenon was first observed by Reuss in 1807 (Reuss, 1809) and is known as electro-osmosis.

A schematic diagram of Reuss' experiment is shown in Figure 2.13. Two glass tubes containing electrodes are introduced into a mass of soil and are filled with water. A direct current is then passed through the soil and after a while it can be observed that the level of the water at the cathode has risen while that at the anode has dropped. At the same time the water surrounding the anode has become cloudy, indicating the presence of some clay particles. The experiment illustrates clearly not only the electro-osmotic flow but also the movement of solid particles, possessing an electrokinetic potential, towards the electrode of the opposite sign. The latter phenomenon is known as electrophoresis.

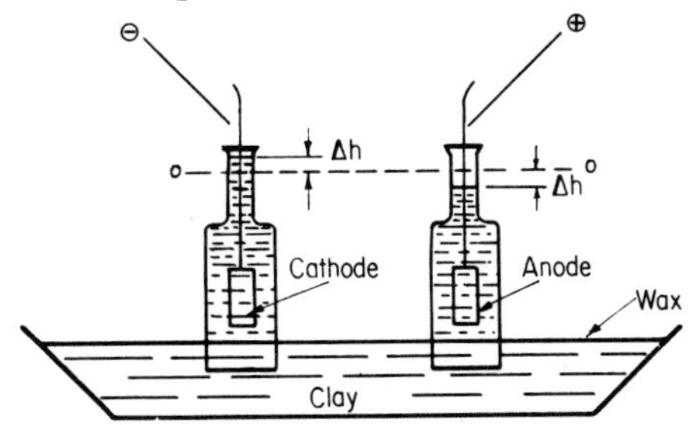

Figure 2.13. Details of Reuss' experiment.

Quincke (1861) discovered that, by passing distilled water through a soil, potential difference is induced between two electrodes placed in the soil in line with the flow:

for silica sand, 6·9 V
for a clay, 0·4 V

This phenomenon, opposite to electro-osmosis, is known as the flow potential.

The electrokinetic phenomena have several practical applications, the most frequently used being electro-osmosis. Its first application was in dewatering an excavation during the construction of a railway line near Saltzgitter. A long,

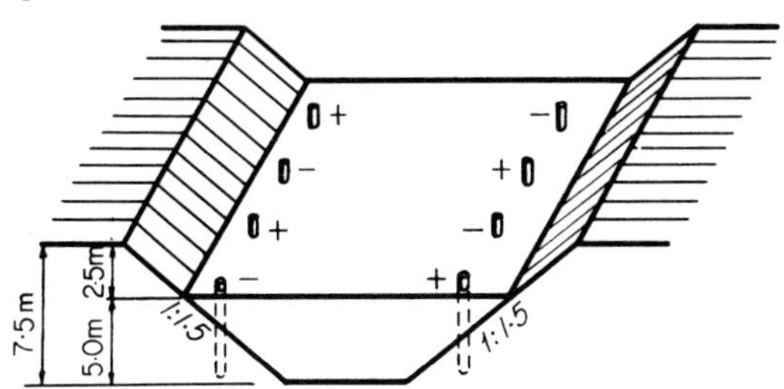

Figure 2.14. Schematic arrangement of cathode wells and anode rods for dewatering of soils by electro-osmosis.

deep cutting had to be made through waterlogged (saturated) silts and at approximately one-third of the required depth the earth works were stopped by continuous collapse of the side slopes. This was so serious that consideration was given to realignment of the line in spite of the fact that the approaches on both sides of the cutting had already been constructed. The situation was saved by application of electro-osmosis to dewatering of the silts, steel rods being used as anodes with cathodes made of perforated, 100 mm (4 in) diameter, aluminium tubes, which also acted as collecting wells (Figure 2.14). Until the current was switched on the yield of the wells was negligible because of the low permeability of the silts. On application of the electric current (90 V d.c.) the combined yield of twenty wells rapidly increased to 2·5 m^3/h (550 gal/h). After several days the soil was sufficiently dewatered to permit excavation with almost vertical side slopes. The earth-moving operations were resumed and continued without any further difficulties. Although the water content of the soil was only decreased from 23·1% to 17·3%, the reasons for the success became obvious when the limits of consistency of the soil were considered: $w_l = 21{\cdot}8\%$ and $w_p = 19{\cdot}2\%$. Owing to lowering of the water content the soil has changed from a liquid to a semi-solid state.

Also the effects of electro-osmosis are used in the introduction of electrolytes into soils to strengthen them or to seal the pores. This method was first used in Switzerland and was later adopted by Cerbertowicz (Zielinski, 1956) to multiple injection of sodium silicate (water glass) and calcium chloride for strengthening and sealing sands.

Electro-osmosis ensures penetration of the solution into pores partly filled with the gelled silica from the first injection. Without application of the electric current it would be necessary to use very high injection pressures; a prolonged application of the electric current accelerates the hardening process of the silica gel.

Further, there are known cases of application of electro-osmosis to the injection of solutions of calcium chloride for strengthening of saturated clays below railway tracks (Zinkin, 1956); the introduced solution of calcium chloride coagulates the clay particles and the surplus water is collected in cathode wells from which it is removed by pumping.

As can be concluded from laboratory and field tests the electro-osmosis gives good results in dewatering silty and clayey soils which have a low permeability. On application of an electric current, with a potential drop of 0·1 V/mm a rate of flow of water of 5×10^{-4} mm/s (5×10^{-5} cm/s) can be achieved (Casagrande, 1949), i.e. between 10 to 10 000 times more than that due to the hydraulic flow. Sands cannot be dewatered with this method because the permeability coefficient of these soils is much greater than 5×10^{-4} mm/s.

2.4. Thixotropy

It is well known that, after standing for a while, certain suspensions of very fine particles (colloids) set to a jelly-like substance—gel. If it is disturbed by agitation or vibration, a gel can be reconverted into a liquid suspension (sol). This reversible phenomenon of re-conversion of gel into sol, and vice versa, due only to mechanical agitation is known as thixotropy.

Thixotropy differs from coagulation in that during coagulation separate flocs form with no bonds between them; whereas in conversion into gel all particles of a given suspension are involved and form a continuous framework like that of a house of cards (Figure 2.15).

Soils containing very small clay particles of colloidal size (< 0·0002 mm) also exhibit thixotropy. It can even be exhibited by soils containing particles much larger than the colloids, e.g. particles of silt or fine sand. By forming thixotropic

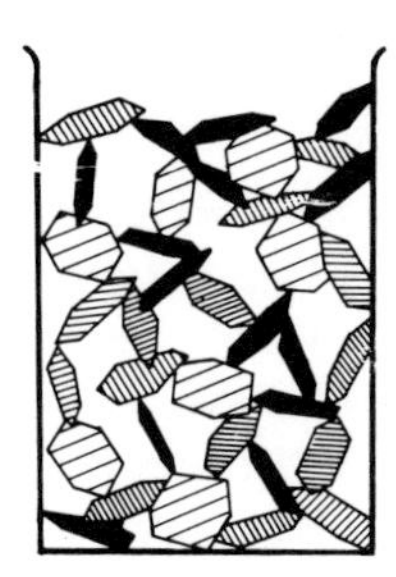

Figure 2.15. Thixotropic structure of clay particles.

links between the larger grains, clay and colloidal particles give the soil apparent cohesion and strength. However, such a thixotropic soil structure can be disturbed by a powerful shock or vibration reducing its consistency to plastic or even leading to its liquefaction.

Such phenomena were observed on several occasions by Wiłun during mechanical earth-moving operations in holocene sandy loams in Warsaw. Transmission of engine vibrations to the soil through the tracks of excavators and graders gradually weakened the structure of the soil which eventually began to wave in front of the lightly loaded tracks. It was found that in all cases the soil was plasticized (i.e. its structure has been disturbed) to a depth of approximately 0·4 m (16 in). On removal of this layer the material was capable of taking a safe bearing stress in the region of 250 kN/m^2 (approximately 2·5 kgf/cm^2, 5050 lbf/ft^2).

Bentonite is a classical example of a thixotropic material which for a long time has been used as a binding agent in foundry sands and for forming water seals in granular materials in the construction of dams and barrages. More recently bentonite has been used extensively as a thixotropic suspension in the construction of diaphragm-retaining cut-off walls in granular soils; its thixotropic properties help to retain the sides of the excavation during the construction of these walls. Several methods of activation of ordinary clays to increase their thixotropic properties have also been developed (Piaskowski, 1956).

In spite of the wide use of thixotropic materials the phenomenon itself is not clearly understood. The most probable hypothesis suggests that at certain concentrations the suspended clay particles are free to undergo rotational movements under impulses received from the free-moving water molecules (thermal movements). They may also be close enough for the attractive forces between the positively charged particle edges and the negatively charged flat faces to take over and to form a house of cards type of framework–gel. The water molecules are then enclosed within the interstices of this continuous framework so that they are no longer freely mobile. A powerful shock will break down the weak electrical bonds between the particles and hence the framework and the water will be free again to undergo thermal movements.

2.5. Capillarity

Continuous channels of pores in soils can be considered as capillary tubes. It is well known that, when a capillary tube is placed in water, the water will rise inside the tube to a certain level above its surrounding free surface. The height of the capillary rise depends on the diameter of the tube; the smaller the tube, the higher the level to which the water rises inside it.

Capillarity results from a combined action of two phenomena: (a) attraction (adhesion) of water to tube walls, and (b) water surface tension.

On immersion of a capillary tube in water its outer and inner surface become coated (owing to molecular attraction) with a thin film of adsorbed water which

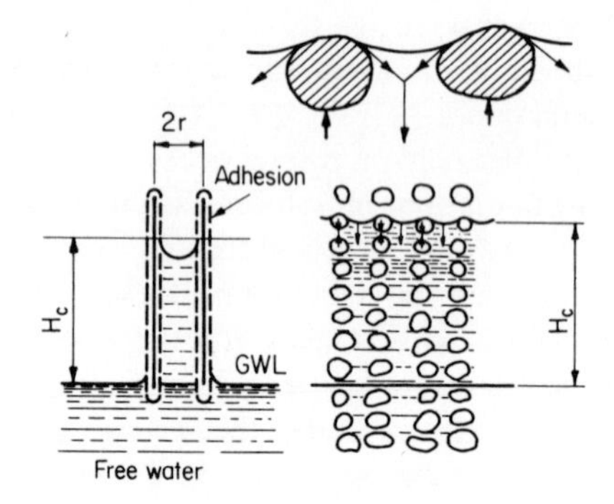

Figure 2.16. Capillary rise of water in glass tube and in soil.

extends to a certain height above the free surface (Figure 2.16). This leads to an increase in the area of the interface between water and air.

The increase of the interface area is opposed by the surface tension. Because of the considerable difference between the density of air and water the cohesion between the water molecules is much stronger than the adhesion between the water and air. This produces a state of tension between the water molecules within the water surface and makes it behave as though it was covered with a thin stretched rubber membrane which continually tends to diminish its surface area.

The water in a capillary tube behaves in a similar manner, rising as long as the weight of the water column does not balance the resultant surface tension σ_{st} acting within the interface, i.e. around the perimeter of the tube. The height of the capillary column H_c above the free surface of water can be evaluated (Figure 2.16) as follows.

The weight of the column of water is

$$W = H_c \pi r^2 \rho_w g^*$$

and the resultant of surface tension

$$Q = 2\pi r \sigma_{st}$$

Equating the right-hand sides of both equations

$$H_c = \frac{2\sigma_{st}}{r\rho_w g} \tag{2.2}$$

Taking into consideration that the surface tension σ_{st} at 10 °C is equal to 0·073 N/m (0·0742 gf/cm) and assuming that the density of water $\rho_w = 1{\cdot}0$ g/ml a simplified formula for the capillary rise of water is obtained:

$$H_c = 15/r \text{ (mm)} \tag{2.3}$$

where r = radius of capillary tube (mm).

Substituting $r = 0{\cdot}1$ mm into the above equation $H_c = 150$ mm is obtained which correlates well with the observed capillary rise of water in a sand with grain sizes between about 0·2 and 0·5 mm. In the case of clays the pore diameters are in the region of 0·1 μm or even less; corresponding calculated capillary rise H_c would be equal to 150 m. Observed heights do not, however, exceed 3 to 4 m; this can be explained by the fact that in the very small capillaries their entire cross-sectional areas are occupied by the strongly held adsorbed water which prevents the upward movement of free water.

The rise of water in the capillaries with relation to the free water level described above is referred to as the active capillarity (Figure 2.17(a)).

Lowering of the free water level with relation to the level of the water in the capillaries is referred to as passive capillarity (Figure 2.17(b)). In such cases the difference of height between the menisci in the capillaries and the free water level can be very high (in accordance with Equation (2.3)). The maximum possible difference of levels between the menisci of the capillary water and free water level is known as the passive capillary rise H_{cp}.

The phenomenon of the capillary rise of water and the fact that it remains suspended above the free water level suggests that the capillary water must be in tension (suction) and the soil skeleton in compression (Figure 2.18). The magnitude of the compressive stress in the soil skeleton can be obtained from the following equation:

$$\sigma'_c = H_c \rho_w g \tag{2.4}$$

where H_c = height of capillary column above free water level
ρ_w = density of water
g = gravitational acceleration

* g, the gravitational acceleration is only used in the S.I. system because ρ_w represents the mass of water per unit volume.

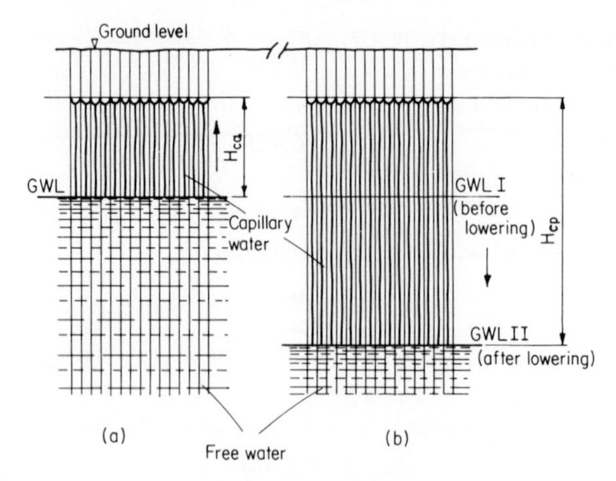

Figure 2.17. Capillarity: (a) active; (b) passive ($H_{ca} < H_{cp}$).

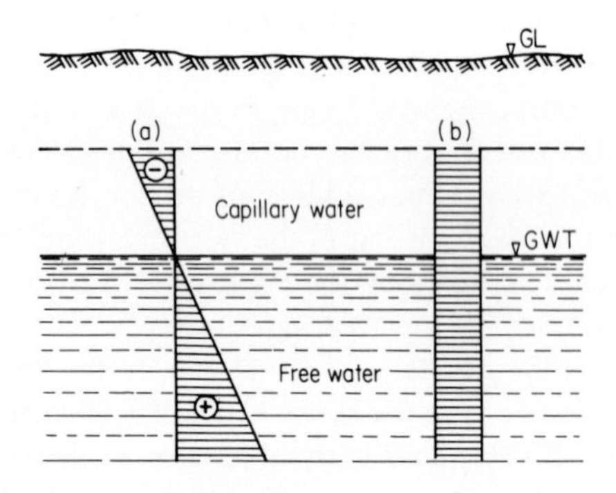

Figure 2.18. Distribution of pressure in water and in soil skeleton: (a) minus sign indicates lower pressure than atmospheric in capillary water, plus sign indicates pressure above atmospheric; (b) stresses in soil skeleton due to suspended capillary water.

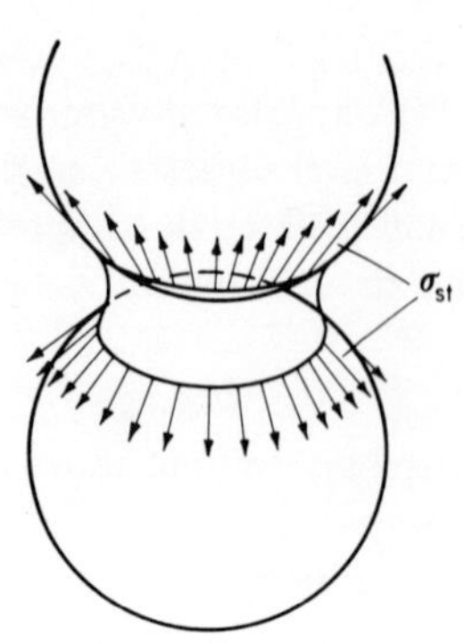

Figure 2.19. Contact stress between two particles induced by local meniscus.

It must be pointed out that the effective compressive stresses induced in the soil skeleton by the capillary forces are uniformly distributed not only throughout the capillary zone but also below the free water level (Figure 2.18(b)).

In addition to the above-described classical cases the phenomenon of capillarity is of considerable importance in partly saturated soils, where compressive contact stresses between individual particles are induced by local menisci (Figure 2.19). The surface tension in the local meniscus induces suction in the pore water which compresses the particles together (Figure 2.20). The magnitude

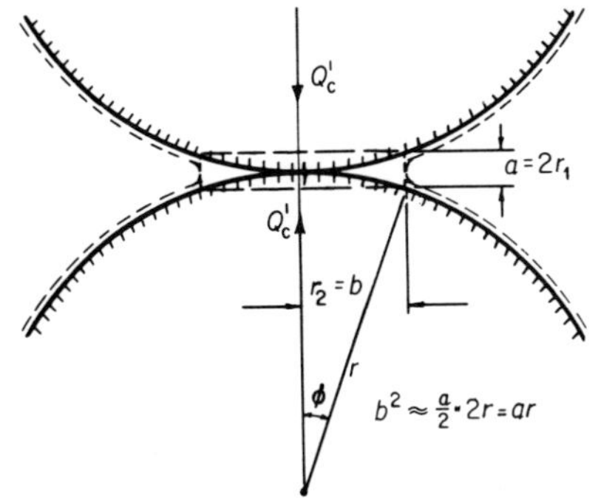

Figure 2.20. Section through two spherical particles and the local meniscus.

of the resultant contact force Q_c' can be computed according to Laplace's equation, taking into consideration the double curvature radii $r_1 = a/2$ and $r_2 = b$ (Bernatzik, 1947):

$$Q_c' = \pi r_2^2 \sigma_{st}\left(\frac{1}{r_1} - \frac{1}{r_2}\right) = \pi b^2 \sigma_{st}\left(\frac{2}{a} - \frac{1}{b}\right)$$

$$\approx \pi b^2 \sigma_{st} \frac{2}{a} \approx 2\pi\sigma_{st}\, r\frac{a}{a} = 2\pi\sigma_{st} r \quad (2.5)$$

It follows from Equation (2.5) that the force is independent of the size of the meniscus and for any given particle diameter can be considered constant. In a soil skeleton consisting of equal size grains the average compressive stress σ_c' can be taken as

$$\sigma_c' \approx \frac{Q_c'}{(2r)^2} = \frac{\pi\sigma_{st}}{2r} \quad (2.6)$$

For $r = 0{\cdot}1$ mm (as for medium sand), $\sigma_c' = 1{\cdot}2$ kN/m² (0·17 bf/in², 0·012 kgf/cm²), and, for $r = 0{\cdot}1$ µm (as for clay), $\sigma_c' = 1{\cdot}17$ MN/m² (170 lbf/in², 12 kgf/cm²). As can be seen, the smaller the radius of given particles the greater is the contact stress between them.

A classical example of the action of capillary forces is observed on a beach of loose sand; when moist (partly saturated), the sand exhibits a considerable strength and easily supports a man's weight but, when it is submerged by a wave, the capillary forces are destroyed, the sand loses its strength, and his foot sinks into it.

2.6. Influence of Physico-chemical Characteristics on Physical and Mechanical Properties of Soil

2.6.1. INTERPARTICLE FORCES IN SOILS

The phenomena discussed in the preceding sections show clearly that the interparticle forces can be divided into two categories.

(a) Attractive (compressive) forces σ_a' which are principally due to molecular attraction between particles (van der Waals' forces).

(b) Repulsive forces σ_r' which are basically electrostatic and are due to the presence of the electrokinetic potentials of equal sign in the diffuse layer

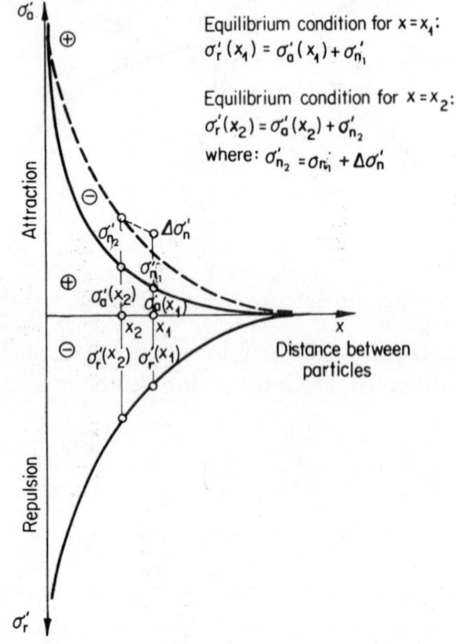

Figure 2.21. Variation of magnitude of interparticle forces with relation to the distance between the particles.

surrounding the soil particles, supplemented by the repulsion between the oriented water diapoles in the adsorbed water layers of adjoining particles.*

The attractive forces may be supplemented by compressive forces σ_n' due to external loading.

The magnitude of these forces in relation to the distance between the particles can be illustrated as shown in Figure 2.21. When particles are very close together the mutually attractive forces are much greater than the repulsive

* Water diapoles in the adsorbed water layer have their negative poles directed either towards the adsorbed layer of cations or towards individual cations in the diffuse layer; this leads to some repulsion between positive poles of water molecules facing each other in the adsorbed water layers of two neighbouring particles and hence to repulsion between the particles.

forces, and permanent bonds between particles may result. As the spacing increases the attractive forces diminish and the repulsion due to the wedging action of the adsorbed water becomes greater than the attractive forces, and the particles will move apart until equilibrium is reached. The action of the repulsive forces can be observed in the resistance to compression during consolidation of soils and in the compressive strain recovery on unloading of soils.

When additional external loading is applied the resultant of the compressive forces between soil particles becomes greater than the repulsive forces and the particles move closer together until equilibrium is achieved (Figure 2.21) or until the particles are touching.*

In spite of the fact that in the condition of equilibrium the sum of the resultant compressive and repulsive forces acting between two neighbouring particles is equal to zero, the existing forces offer frictional resistance to relative translation (shearing) of the particles. The resistance to shearing which is solely due to the attractive forces between the particles is known as the true cohesion of soils; it is denoted by symbol c_e' and is expressed in terms of a force per unit area.

The resistance due to the compressive interparticle forces induced by the external loading is taken as the frictional resistance, equal to the product of the normal stress σ_n' and the coefficient of friction.

The total shearing resistance can be expressed as the sum of the two components as follows:

$$\tau_f = c_e' + \sigma_n' \tan \phi_e' \tag{2.7}$$

In accordance with Figure 2.21 the true cohesive resistance c_e' can be expressed as

$$c_e' = \sigma_a' \tan \phi_a' \tag{2.8}$$

It follows from the above that the true cohesion c_e' is a function of the distance between the particles.

Soils consist of a large number of individual particles. The smaller they are, the greater is their number, the greater is the number of contacts per unit volume, and the greater is the cohesion exhibited. The effect predominates in clays which are consequently known as cohesive soils. Since the number of particle contacts per unit volume also increases with density of packing, it follows that the cohesion increases with consolidation and expulsion of water from the soil.

Coarse-grained soils have no cohesion, the gravitational forces on individual particles being many times greater than the attractive forces, and the number of contacts (per unit volume) is small; these are known as cohesionless soils (gravels and sands).

* Soil surface activities have not yet been fully explained; the above description of interaction between soil particles is given here by the authors as hypothetical for pictorial explanation of the basic behaviour of soils.

The shearing resistance of a soil depends on both the cohesion and the internal frictional resistance between particles. In cohesive soils, shearing resistance is predominantly produced by cohesion, whereas in cohesionless soils internal friction predominates.

The cohesive resistance of soils may also result from cementation of the points of contact between particles by crystallization of salts in the pores. Cementing action is particularly noticeable in dried-out soils, the increased concentration of salts in the pore fluid assisting their crystallization.

Other factors such as prolonged consolidation at very high pressure and temperature changes can, under favourable conditions, increase the cohesion of soils by formation of the diagenetic bonds (Bjerrum, 1967). The nature of these bonds has not yet been fully explained but their existence in over-consolidated soils is indisputable.

Cohesion of fine-grained soils is also influenced and can be increased by surface tension in the local capillary menisci. Such an increase in cohesionless soils is only apparent and is lost on disappearance of the menisci (drying-out of sand or its submergence).

2.6.2. SHRINKAGE AND SWELLING OF COHESIVE SOILS

If a small lump of a cohesive soil is desiccated, it can be observed that its volume decreases. A decrease in volume of a soil can only take place under the action of external compressive forces with a free drainage of the water contained in the pores. Both the above phenomena take place during the desiccation process of a cohesive soil.

Owing to evaporation of water from the surface of the soil, the capillary menisci retreat into the narrowing pores exerting an increasing capillary pressure which is accompanied by increasing interparticle stresses and the reduction in the thickness of the adsorbed water films at the points of contact; this leads to a decrease in volume of the drying lump of soil.

The decrease in volume of a soil during desiccation is known as the shrinkage.

In the case of wetting of a desiccated soil a phenomenon opposite to the above described takes place, that is to say the soil swells. The water reaching the menisci on the surface of the soil decreases the capillary pressure which leads to the reduction of the interparticle contact stresses.

The adsorbed water films at the points of contact tend to achieve the complete hydration of the adsorbed cations by drawing water molecules back from the free water present in the pores. The place of the molecules drawn into the adsorbed layers is taken up by the water drawn in from the surface of the soil. The volume of the soil increases, i.e. the soil swells. At the same time the water content of the soil increases while the pore water suction and the soil strength decrease.

2.6.3. EQUILIBRIUM OF WATER CONTENT IN SOILS

The changes that simultaneously occur in the water content of the soil and in the pore water suction can be experimentally investigated during both the desiccation and wetting processes. It has been established that for any soil in the condition of complete saturation there is a definite relationship between magnitude of the pore water suction and the water content. According to Schofield (Road Research Laboratory, 1955) this relationship is different for desiccation and wetting of the soils (Figure 2.22). The investigation involves

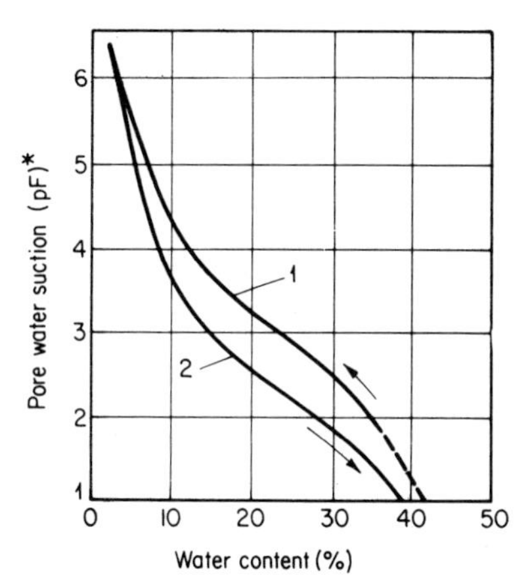

Figure 2.22. Relationship between pore water suction and water content: 1, on desiccation; 2, on wetting (after Schofield, 1955; Crown copyright, reproduced by permission of the Director of Road Research).
* On the *p*F scale of measuring suction, the logarithm to base ten of the suction expressed in cm of water is equivalent to the *p*F value (1 cm of water = 0·98 kN/m^2).

measurement of the pore water suction with the help of a manometer. The results of the tests carried out in the British Road Research Laboratory are given in Figure 2.23; as can be seen from the results the more clayey the soil the greater is the pore water suction at the same water content.

At the same time it is important to note the fact that at low water contents the pore water suctions observed are very high, while in the case of complete saturation they are negligibly small. The water content of a soil of natural structure at which, on complete saturation, the pore water suction is equal to zero, can be referred to as the absorption limit, w_a. Investigations can be carried out using Enslin's continuous-flow apparatus (Figure 2.24) (Kézdi, 1964).

The experimental procedure using Enslin's apparatus is as follows. A dry powdered sample of the soil is placed on the suction plate (high air entry value porous disc) and the tap connecting the water in the calibrated tube with that below the plate is opened. The suction in the soil draws the water from the tube. The ratio of the mass of the absorbed water to the mass of the dry soil (in percentages) defines the absorption of the soil.

The relationship between the pore water suction and water content in soils (Figure 2.22) is the result of and exists only in the presence of capillary pressures. In these conditions the pore water suction is resisted and balanced by compressive stresses within the soil skeleton. Hence it can be stated that the same relationships exist between the stresses within the soil skeleton and water

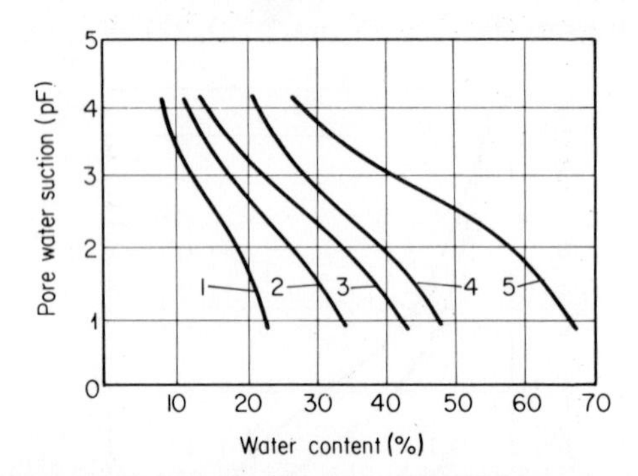

Figure 2.23. Relationship between pore water suction and water content for different soils: 1, Culham sand; 2, Harmondsworth sandy clay; 3, Pottery clay; 4, Norton clay; 5, London clay (after Schofield, 1955; Crown copyright, reproduced by permission of the Director of Road Research).

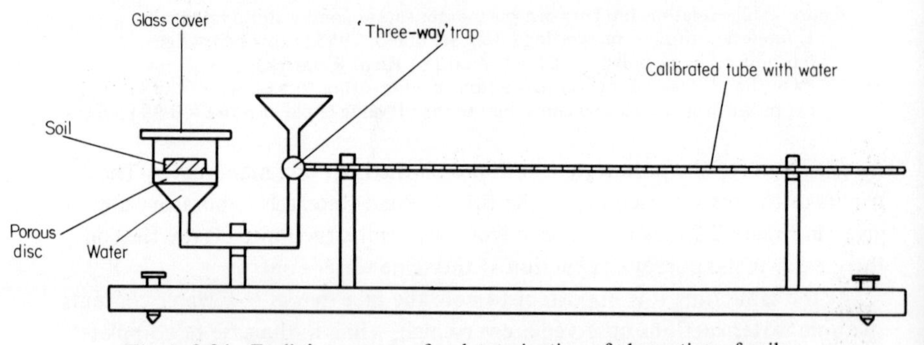

Figure 2.24. Enslin's apparatus for determination of absorption of soils.

content as between the pore water suction and water content (obviously only in the case of exclusive action of capillary pressures).

If at a certain level the soil layer is under the combined action of stresses induced by external loading, self-weight of the overlying soil, and the capillary pressure, then the interparticle stress will be equal to the sum of the applied stresses:

$$\sigma'_{iz} = \sigma'_z + \sigma'_{oz} + \sigma'_c \tag{2.9}$$

where σ'_{iz} = effective interparticle stress in kN/m^2, determined as the force acting between the particles over a unit cross-section area of the soil (area of particles and pores)
σ'_z = effective stress component induced by external loading in kN/m^2
σ'_{oz} = effective overburden stress in kN/m^2
σ'_c = capillary pressure determined according to Figure 2.18 in kN/m^2

Having obtained the effective stress within the soil skeleton at a given level it is now possible, from a pore water suction–water content relationship similar to that shown in Figure 2.22, to determine the water content of the soil at that level.

The relationship between the effective stress and water content of the soil can also be evaluated directly from the oedometer test (see Section 5.2). This test can also be used for determination of the relationship between the pore water suction in the soil and its water content but this will only be valid for the *in situ* conditions in which the effective stresses induced in the soil by the pore water suction and loading are the same as the stresses induced in the oedometer.

It must be mentioned, for the sake of consistency, that the effective stress in Equation (2.9) is the vertical component of the stresses induced in the soil skeleton at the given level and that the horizontal stress components σ'_x and σ'_y are usually smaller. Only in the investigation of the pore water suction in an isolated lump of soil are all the effective stress components equal ($\sigma'_z = \sigma'_x = \sigma'_y$). In connection with this the relationship between the water content and effective stresses in the soil should be considered in terms of the effective mean stress $\sigma'_m = \frac{1}{3}(\sigma'_z + \sigma'_x + \sigma'_y)$ or, in terms of the vertical stress according to the oedometer test, if this mode of loading corresponds closely to the *in situ* conditions beneath a road surface or a foundation.

It follows from the above considerations that in all cohesive soils there is a definite relationship between the effective stresses and the corresponding water contents of the soil. This relationship can be defined as the equilibrium of water content. It is strictly connected with magnitude of the repulsive and attractive forces acting between the soil particles (Figure 2.21).

The equilibrium of water content can be expressed as

$$\sigma'_r = \sigma'_a + \sigma'_n \qquad (2.10)$$

where σ'_r = repulsive force between adjacent particles
σ'_a = attractive force
σ'_n = effective interparticle force representing combined action of external loading, overburden stresses, and capillary pressure

The magnitudes of σ'_r and σ'_a depend on the distance between the particles (Figure 2.21). Determination of the absolute value of these forces is not possible at present and therefore we use their difference, which is equal to σ'_n.

Figure 2.21 shows schematically the influence of an increase in external loading on the distance between the particles and on the internal forces.

If the forces acting on two particles, distance x_1 apart, are in equilibrium then

$$\sigma'_r(x_1) = \sigma'_a(x_1) + \sigma'_{n_1}$$

On application of an additional loading $\Delta\sigma'_n$ there is an out-of-balance compressive force which induces the particles to move closer together until, at a distance x_2 apart, the equilibrium is again reached:

$$\sigma'_r(x_2) = \sigma'_a(x_2) + \sigma'_{n_2}$$

2.6.4. COMPRESSIBILITY AND EXPANSION (RECOVERY) OF SOILS

Under the action of external compressive loading the soil decreases in volume. This property of the soil is known as the compressibility.

When the soil is unloaded its volume increases; this phenomenon is known as expansion or recovery (often referred to as swelling). The expansion should not be confused with the true swelling, also involving an increase in volume which, however, takes place not as the result of decreasing loading but owing to the absorption of water under a constant external load.

The mechanism of compressibility and expansion is related to the equilibrium of water content in the soil. The phenomenon of compressibility is accompanied not only by the change in water content but also by the decrease in the distance between the soil particles and hence by the change in the magnitude of the repulsive and attractive forces between the particles.

3

Physical Properties of Soils and Laboratory Methods of their Determination

3.1. Phases in Soil

Soil consists of individual grains and particles which form a porous body. Depending on the conditions under which the soil has been deposited and on the loading history and water content the pores in the soil are filled either with water or with air or, as most frequently is the case, with both these substances together.

The grains and particles form the solid phase of the soil, water forms the liquid phase, and air (water vapour or gases) the gaseous phase.

Proportions by volume expressed in percentages of individual phases in a given soil can be illustrated by a single point on a graph with triangular co-ordinates (Figure 3.1). The changes occurring in the volume of different

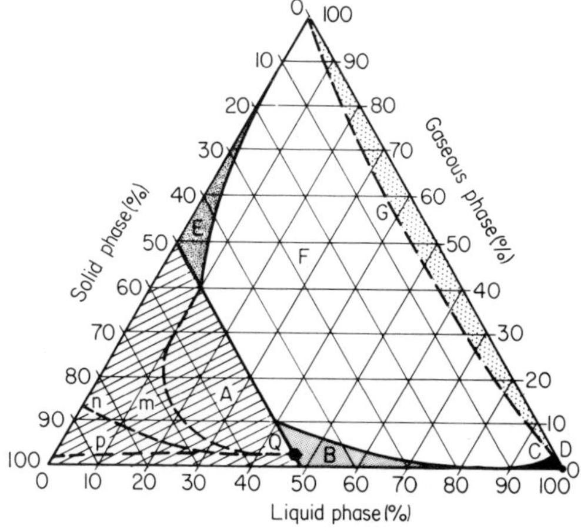

Figure 3.1. Phases in soil: A, composition of natural soils; B, alluvium mud, recent sediments of silts or clays in water; C, soil suspensions; D, water; E, wind-blown silts; F, foamy soil suspension; G, rain with dust.

phases of a given soil Q during compaction, desiccation, or consolidation are illustrated in the figure by dashed lines m, n, and p respectively; at any instant during the testing the volume of soil is taken as equal to 100%.

3.2. Soil Structure

The arrangement of soil grains and particles forming the soil skeleton is known as the *soil structure*. The soil structure depends on the size and type of particles present and on the conditions under which the soil has been formed. The main types of soil structure are the following: single grain, honeycomb, and flocculent (Figure 3.2).

Single grain structure (Figure 3.2(a)) is characteristic of sands and gravels in which the attraction forces between grains are negligible; porosity typical of this type of structure varies from 20 to 50%.

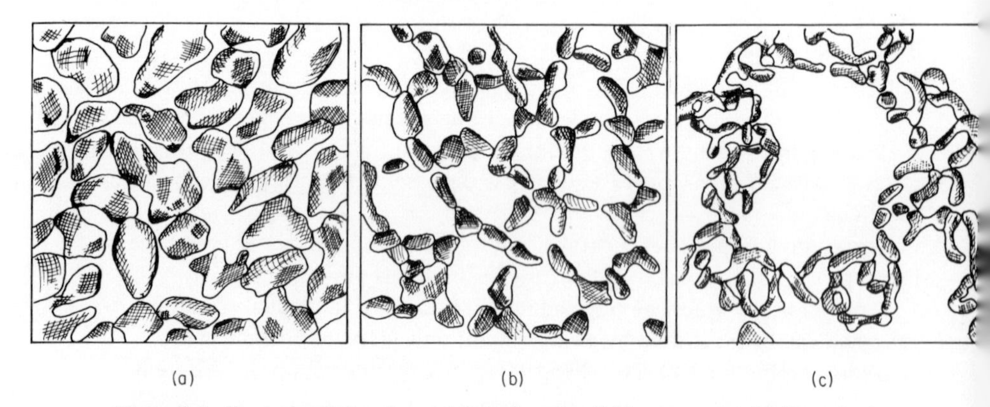

Figure 3.2. Typical soil structures; (a) single grain; (b) honeycomb; (c) flocculent.

Porosity depends on the conditions under which a given soil has been deposited; soils deposited by slowly flowing waters or by wind possess a relatively large porosity, while those deposited by fast flowing waters are much denser.

Honeycomb structure (Figure 3.2(b)) is characteristic of clayey soils deposited in waters without prior coagulation of falling particles. Its formation is illustrated in Figure 3.3. The falling clay and silt particles are, on contact with the already deposited particles, attracted to them with forces greater than their weight and, therefore, form a skeleton of a honeycomb-type structure. Soils which have been formed in this manner have a porosity much in excess of 50%.

Flocculent structure (Figure 3.2(c)) is almost solely formed by clay particles sedimenting in waters rich in dissolved salts. These solutions reduce the electrokinetic potential of the particles which induces their coagulation while they are still in suspension; on sedimentation a honeycomb-type structure is formed, the

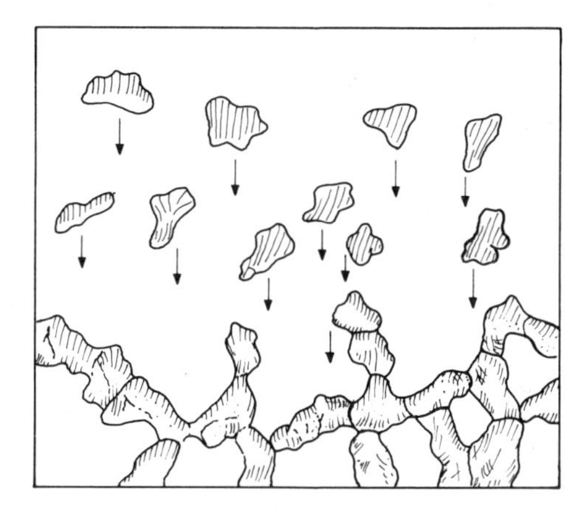

Figure 3.3. Formation of honeycomb structure.

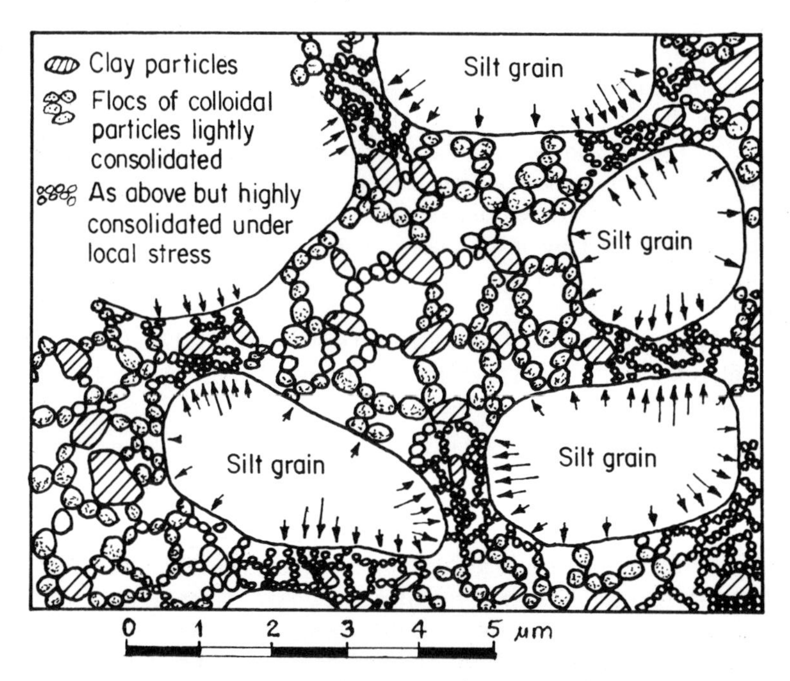

Figure 3.4. Mixed structure (after Casagrande).

skeleton of which consists of the primary floc cells. The porosity of this type of structure is very high.

Very frequently natural soils possess a mixed structure as shown in Figure 3.4.

3.3. Particle-size Distribution in Soils

Particle-size distribution expresses the size of the particles comprising a given soil in terms of percentages by weight of individual sizes. In practice five basic grain and particle* sizes are used as shown in Table 3.1.

Table 3.1. Basic particle sizes (based on British Standard 1377 (1967))

Name of range of particle size	Range of particle size
stone boulders	> 200 mm
cobbles	200–60 mm
gravel	60–2 mm
sand	2–0·06 mm
silt	0·06–0·002 mm
clay	< 0·002 mm (2 μm)

In practice natural soils contain only a small proportion of grains greater than 2 mm. Therefore, in order to describe a soil in terms of its particle-size distribution a classification has been developed which is based only on the smallest three particle sizes: sand, silt, and clay. For illustration of the relative contents of these particle sizes Feret's triangle can be used as shown in Figure 3.5. Two versions of the classification of the soils within the triangle are given: the first one uses the term loam (a mixture of sand, silt, and clay in varying but approximately equal amounts) while the second recommended by the authors avoids the use of this rather ambiguous term and defines the soils in terms of the basic types only.

Example. Given the results of a particle-size distribution analysis (sieve-sedimentation), define the type of the soil according to the two classifications shown in Figure 3.5:

sand content, 67%
silt content, 27%
clay content, 6%

On the left side of the triangle a point corresponding to 67% sand content is marked. Similarly on the base of the triangle 27% silt content is marked. Two lines are now drawn through these points parallel to the sides of the triangle. The name of the zone in which the point of intersection M of these lines falls indicates the type of the soil: 'sandy loam' according to the U.S. Bureau of Soils' classification and 'slightly clayey sand' according to the classification suggested by the authors.

* When grains and particles are mentioned in the text the former are taken to be rounded, usually, and visible to the naked eye and of dimensions greater than 0·06 mm, while the latter have dimensions smaller than 0·06 mm.

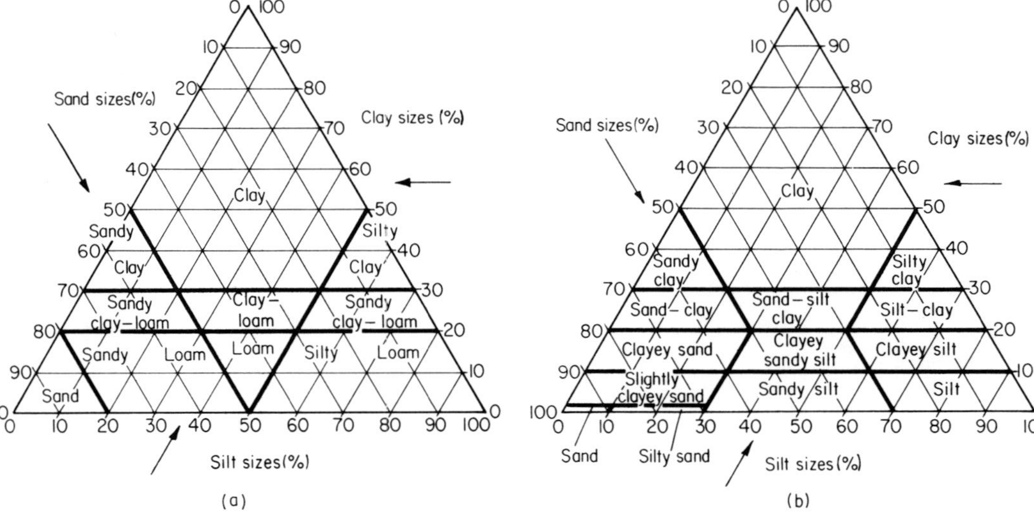

Figure 3.5. Classification of fine-grained soils by means of Feret's triangle: (a) U.S. Bureau of Soils and Public Road Administration; (b) based on the Polish Standard PN-54/B-02480 (1954).

Apart from the above classification of the cohesive soils in terms of the three particle-size ranges an additional classification of cohesionless soils (gravels, sands, and silts) as shown in Table 3.2 can be made and is of significance in civil engineering (Glossop and Skempton, 1945).

Table 3.2. Classification of cohesionless soils

Type name	Grain content	
	In size (mm)	In percentages
gravel: coarse	60 – 20	over 50% of particles within the specified size limits but, for gravels and sands, silt content less than 30% and clay content less than 2% while, for silts, clay content less than about 10%
medium	20 – 6	
fine	6 – 2	
sand: coarse	2 – 0·6	
medium	0·6 – 0·2	
fine	0·2 – 0·06	
silt: coarse	0·06 – 0·02	
medium	0·02 – 0·006	
fine	0·06 – 0·002	

3.4. Determination of Particle-size Distribution in Soils

Determination of particle-size distribution (granulometric analysis) in soils uses the following three methods.

(a) Sieving (wet or dry)—for gravels and sands with grains $\geqslant 0{\cdot}075$ mm.

(b) Sedimentation (pipette)—for cohesive soils containing a considerable proportion of particles smaller than 0·075 mm.

(c) Sedimentation (hydrometer)—as above.

For soils containing a wide range of particle sizes a combination of the sieve and sedimentation analysis is necessary.

3.4.1. SIEVE ANALYSIS

In this analysis a dry sample of gravel or sand is passed through a series of sieves of known-aperture (B.S. sieves) and the percentage weight of the grains passing through each individual sieve is computed as follows:

$$s_i = \frac{M_s - M_{ri}}{M_s} \times 100 \tag{3.1}$$

where s_i = percentage weight of grains smaller than d_i mm
M_{si} = dry mass* of soil retained on sieves of aperture $\geqslant d_i$ in g
M_s = total dry mass of the soil sample in g

* The S.I. system of units differentiates clearly between the mass and the gravitational force which is commonly referred to as the weight; gravitational force is equal to the product of the mass and the gravitational acceleration 'g'. In normal laboratory weighing we determine the mass of a body by comparing it with another body of a known mass (standard weights).

The determined values of s_i are now plotted on a particle-size distribution chart (see Section 3.5).

The above analysis should be carried out in accordance with the British Standard 1377 (1967, Test 7A or B).

3.4.2. SEDIMENTATION ANALYSIS

In this type of analysis a suspension of a uniform concentration of soil particles in water is prepared and as the particles sediment the changes in the density ρ_z of the suspension are observed.

The density of the suspension ρ_z depends on the specific gravity of the solid particles present and can be determined from the following expression:

$$\rho_{zi} = M_{si} + \left(\rho_w - \frac{M_{si}}{G_s}\right) = M_{si} + \rho_w - \frac{M_{si}}{G_s} \tag{3.2}$$

where M_{si} = mass of the soil particles contained per unit volume of suspension at a given time t_i in g/ml
ρ_w = density of water at the given temperature in g/ml
G_s = specific gravity of solid particles (dimensionless)

Assuming that the density of water ρ_w = 1·0 g/ml the mass of the solid particles contained per unit volume of the suspension can be computed:

$$M_{si} = \frac{\rho_{zi} - 1}{G_s - 1} G_s \text{ (g/ml)} \tag{3.3}$$

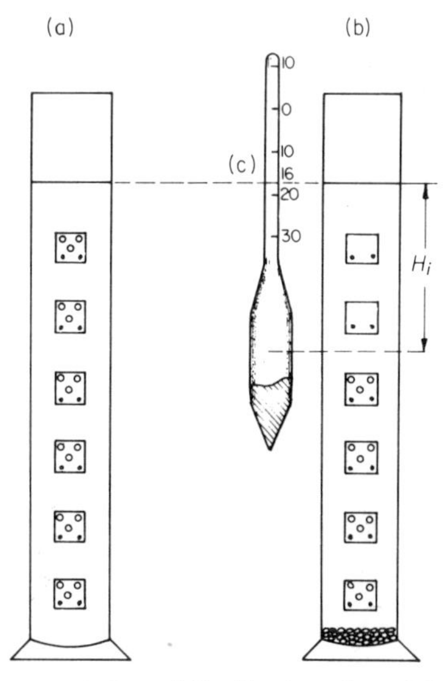

Figure 3.6. Schematic representation of distribution of particles in a suspension; (a) uniform concentration of suspension; (b) change in concentration above depth H_i.

Immediately after a thorough agitation of the suspension the distribution of the solid particles throughout it is uniform and particles of all sizes are present in the same quantity at any point within the cylinder (Figure (3.6a)

As soon, however, as the agitation is stopped sedimentation of the solid particles commences. The sinking velocity of the particles depends on their diameter and on the viscosity of the fluid; according to Stokes' law the velocity of sinking spherical particles in a still fluid is related to these quantities as follows:

$$v_i = \frac{(G_s - 1)\,\rho_w g d_i^2}{180\,\eta} \text{ mm/s (S.I. units)} \tag{3.4}$$

or

$$v_i = \frac{(G_s - 1)\,\rho_w g d_i^2}{18\,\eta} = \frac{(G_s - 1)\,g d_i^2}{18\,\eta} \text{ cm/sec (c.g.s. units)}$$

where η^* = viscosity of water in MN s/mm² (in dyn x sec/cm²)
d_i = diameter of particles in mm (in cm)
ρ_w = density of water in g/ml (in g/cm³ = 1·0)
g = gravitational acceleration in mm/s² (in cm/sec²)
G_s = specific gravity of solid particles (dimensionless)

Particles of the same size sink with the same velocity throughout the whole height of the cylinder. Towards the bottom of the suspension the particles that have sunk lower are being replaced by the same quantity of similar particles arriving from above; therefore initially the density of the suspension at those levels remains constant. However, the density of the suspension in the upper parts of the fluid changes because, after even a relatively short period of time, the larger particles would all have sunk below a certain depth and are no longer being replaced by an inflow of similar particles from above.

Therefore, on elapse of a period of time t_i, particles of a certain diameter d_i, or greater, will not be present in the suspension above the depth H_i; the diameter of these particles can be computed from Stokes' law:

$$d_i = \left\{\frac{180\,\eta}{(G_s - 1)\,\rho_w g} \times \frac{H_i}{t_i}\right\}^{1/2} \text{ mm (in S.I. units)} \tag{3.5}$$

or

$$d_i = \left\{\frac{18\,\eta}{(G_s - 1)\,g} \times \frac{H_i}{t_i}\right\}^{1/2} \text{ cm (in c.g.s. units)}$$

This is accompanied by a decrease in the density of the suspension above the depth H_i.

* For water temperature of 20 °C the coefficient of viscosity η = 0·01 dyn x sec/cm² or η = 0·1 MN x s/mm². For G_s = 2·65 an approximate expression for sedimentation velocity is obtained: $v_i = 9000\,d_i^2$ cm/sec in c.g.s. units and $v_i = 900\,d_i^2$ mm/s in S.I. units.

The variation in time of the density of the suspension at depth H_i can be measured using the pipette or hydrometer method.

(a) *Pipette method.* In this method a pipette is used for drawing off samples of suspension from a specified depth at definite time intervals after the commencement of sedimentation. The samples are usually in 10 ml volume and are taken from a depth of 100 mm.

The mass of the solid particles contained in a sample of volume V_{si} (in ml) is then obtained by evaporation of the water and very accurate weighing of the residual soil M_{si} (in g). The maximum particle diameter d_i in the sample is found from Equation (3.5) by substituting in it the sampling depth H_i and sampling time t_i measured from the commencement of sedimentation. The percentage by weight of particles smaller than d_i is then computed from the following expression:

$$s_i = \frac{M_{si}}{V_{si}} \times \frac{V_{st}}{M_{st}} \times 100 \tag{3.6}$$

where M_{st} = total dry mass of the sample in g
V_{st} = total initial volume of the suspension in ml

Because the pipette method is very tedious and requires very sensitive weighing equipment, the alternative hydrometer method is more commonly used and will now be briefly described.

(b) *Hydrometer method.* The density of suspension ρ_{zi} is measured in this method by means of a hydrometer (Figure 3.6) and, as in the previous method, the readings are taken at definite time intervals t_i after the commencement of sedimentation. The range of densities ρ_{zi} (specific gravities) measured by a hydrometer varies between 0·995 to 1·030 g/ml but for soil-testing purposes the scale is usually calibrated to read R_i from 10 to 30, i.e.

$$\rho_{zi} = \frac{R_i + 1000}{1000} \text{(g/ml)} \tag{3.7}$$

On substitution for ρ_{zi} in Equation (3.3) the mass of the soil particles per unit volume of the suspension is obtained:

$$M_{si} = \frac{(\rho_{zi} - 1)}{(G_s - 1)} G_s = \frac{G_s R_i}{(G_s - 1)\,1000} \text{ (g/ml)} \tag{3.8}$$

It is assumed that the level at which the density is being measured coincides with the centre of gravity of the hydrometer and hence the sampling depth H_i can be measured directly between the point on the scale R_i and the centre of the float.

Given the sampling time t_i and the depth H_i the maximum particle diameter d_i present at the sampling level is found from Equation (3.5). Given the mass of particles per unit volume of suspension M_{si}, the percentage by weight of particles smaller than d_i can be computed from Equation (3.6):

$$s_i = \frac{M_{si}}{1{\cdot}0} \times \frac{V_{st}}{M_{st}} \times 100$$

and if, as is usually the case, $V_{st} = 1000$ ml the expression simplifies to

$$s_i = \frac{G_s R_i}{(G_s - 1) M_{st}} \times 100 \tag{3.9}$$

The above sedimentation analyses should be carried out in accordance with the British Standard 1377 (1967, Tests 7C and D). In the case when it is necessary to determine the coefficient of activity A_c of a soil (see Section 3.9) the particle-size analysis should be carried out on the remains of a sample that was used for determination of the consistency limits; this permits the determination of the correct coefficient of activity for the given soil.

3.5. Particle-size Distribution Curves, Uniformity Coefficient, and Effective Size

The results of a particle-size analysis carried out by one, or by a combination, of the above methods are in the form of a series of percentages by weight s_i of grains or particles of dimensions smaller than certain diameters d_i (fixed by the aperture of the standard sieves or by time intervals in the case of a sedimentation analysis). These results are then usually plotted on a semilogarithmic chart in the form of a particle-size distribution curve: particle diameters d_i are plotted in a logarithmic scale on the abscissa while the percentages s_i are plotted in a linear scale on the ordinate (Figure 3.7).

Particle-size distribution curves are used to determine (a) percentage contents of the basic grain and particle sizes necessary for classification of soils, and (b) particle diameters d_{10} and d_{60} which facilitate the description of certain significant characteristics of soils.

Particle diameter d_{10} is known as the *effective diameter or size*; it is used in determination of, for example, permeability of the soil (Section 4.3).

The ratio of the particle diameters d_{60} to d_{10} is known as the *uniformity coefficient* of the soil U:

$$U = \frac{d_{60}}{d_{10}} \tag{3.10}$$

where d_{60} = particle diameter at which 60% of the soil weight is finer
d_{10} = particle diameter at which 10% of the soil weight is finer, i.e. 90% is coarser than the effective diameter

The particle diameters d_{60} and d_{10} are determined by drawing lines through points 60% and 10% (on the percentage passing scale), parallel to the abscissa,

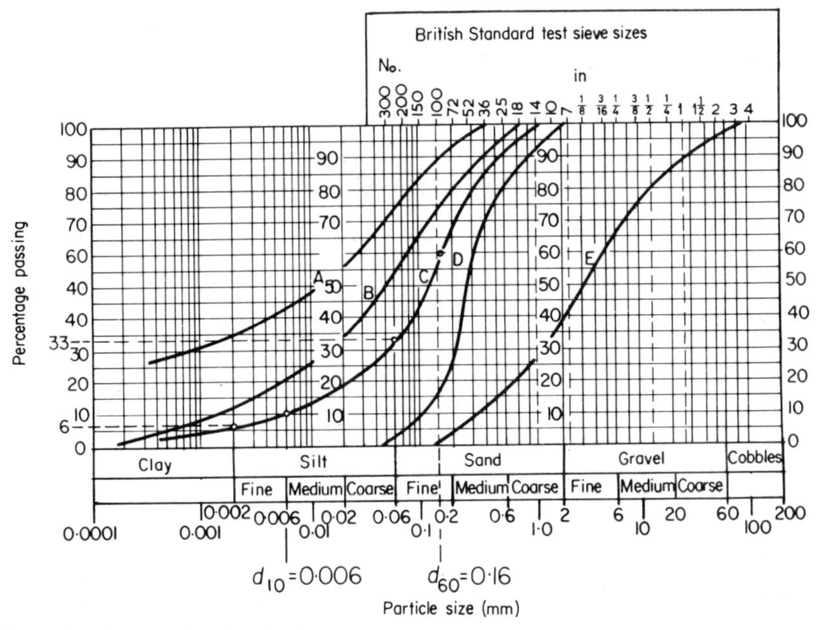

Figure 3.7. Particle-size distribution curves: A, clay; B, clayey sandy silt; C, slightly clayey sand; D, medium sand; E, sandy gravel.

until intersection with the particle-size distribution curve; the projections of these points on the abscissa yield the required results (Figure 3.7).

The value of U defines the uniformity of particle sizes in a given soil (the minimum possible value of U is unity).

When

$U \leqslant 5$, soil is defined as uniform (e.g. dune sands, loess)

$5 < U \leqslant 15$, soil is defined as non-uniform (e.g. holocene silt)

$U > 15$, soil is defined as well graded (e.g. boulder clay)

Example. Determine the percentage contents of the basic particle sizes and the uniformity coefficient of a slightly clayey sand C as shown in Figure 3.7.

Firstly the points of intersection of the curve with the boundaries of the basic particle sizes (size 0·06 and 0·002 mm) are found and then lines (dotted) are drawn parallel to the abscissa to cut the 'percentage passing' scale. The results obtained are 33% and 6% respectively.

Content of sand size (2–0·06 mm) is 100–33 = 67%
Content of silt size (0·06–0·002 mm) is 33–6 = 27%
Content of clay size (< 0·002 mm) is = 6%

Total 100%

Checking the classification of the soil in Figure 3.5(b) we find that the soil has been correctly defined as a slightly clayey sand.

Particle diameters d_{60} and d_{10} are now determined. Horizontal lines are drawn (Figure 3.7) through points 60 and 10% on the 'percentage passing' scale to intersect the particle-size distribution curve of the slightly clayey sand. The points of intersection are projected on the abscissa giving the following results:

$$d_{60} = 0{\cdot}16 \text{ mm and } d_{10} = 0{\cdot}006 \text{ mm}$$

Therefore the uniformity coefficient is

$$U = \frac{0{\cdot}16}{0{\cdot}006} = 27$$

The slightly clayey sand under consideration is well graded, i.e. it contains a very wide range of particle sizes. Its effective diameter d_{10} is on the border line between fine and medium silt sizes.

3.6. Basic Physical Properties of Soils

By the basic physical properties of a soil we mean its water content, natural or bulk density, and density or specific gravity of the solid phase. All these properties are determined from laboratory tests.

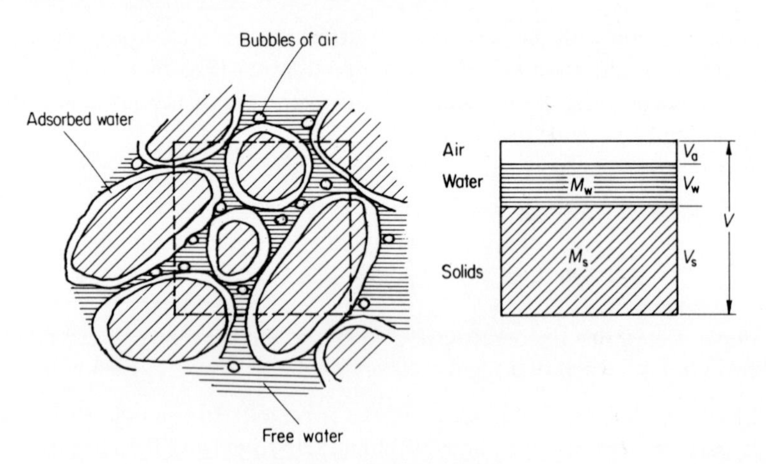

Figure 3.8. Soil constituents.

The knowledge of the basic physical properties is essential for the evaluation of the following additional physical properties: dry density of the soil skeleton, porosity, voids ratio, saturation water content, and degree of saturation.

The knowledge of these basic properties is also essential in the determination of the state of compaction of cohesionless soils by evaluation of the density index, and in the determination of the state of consistency of cohesive soils by evaluation of the consistency index.

A soil is a granular medium consisting of individual grains and particles with the pores usually filled with water containing small bubbles of air (Figure 3.8). The proportions by volume of individual constituents (phases) contained in a given volume of soil can be shown schematically as in Figure 3.8. According to this diagram the following relationships exist:

$$V = V_s + V_w + V_a$$

$$M = M_s + M_w$$

3.6.1. WATER CONTENT

Water content of a soil is the ratio of the mass M_w of the water contained in the pores, to the mass M_s of the solid particles, expressed in percentages

$$w = \frac{M_w}{M_s} \times 100 \tag{3.11}$$

Water present in the pores is in the form of free or gravitational water, capillary water, and adsorbed water; it can also be present as a chemically bonded water (within the crystalline lattice of clay particles).

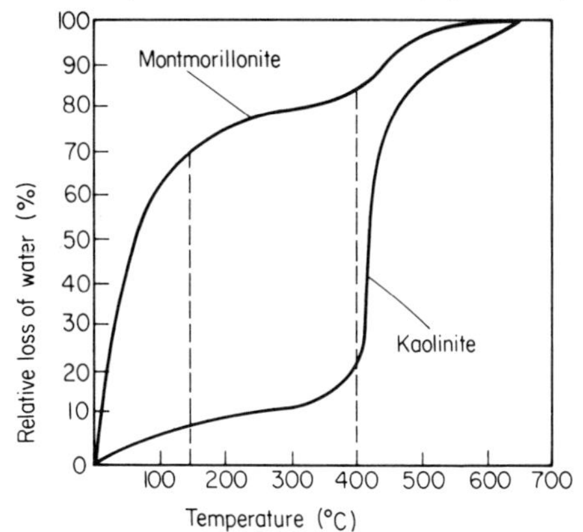

Figure 3.9. Water losses in soil on drying out at increasing temperatures.

If on the drying out of a soil sample a constant mass is reached at a certain temperature level and then the temperature is increased, we shall observe a further decrease in the mass of the soil which, provided the temperature is kept steady at the new value, will again reach a constant minimum value; this can be repeated up to the temperature of +700 °C (Figure 3.9).

On heating up to 150 °C free water, capillary water, and some of the weakly held water on the perfect crystal faces is lost; on further heating from 150 °C to 400 °C the water which is strongly held by the free ions at the points of imperfections in the crystalline lattices and at the edges of the clay crystals is evaporated.

Only on heating to temperatures higher than 400 °C does the chemically bonded water in the form of hydroxides, contained within the crystalline lattices of clay minerals, begin to evaporate.

The curves shown in Figure 3.9 illustrate another very important characteristic of the clay minerals. It can be seen that at 150 °C montmorillonite loses approximately 70% of its water content while kaolinite only loses between 7 and 8%. This implies that the water is only very weakly held by the montmorillonite while it is very strongly held by the kaolinite. In the case of monomineral soils their mineralogical composition can be established by heating them up to high temperatures and observing the water losses.

Because a given soil loses different quantities of water, depending on the temperature to which it is heated, it is necessary in the determination of the water content to dry the soil always at the same constant temperature: in engineering analysis of soils a temperature range of 105–110 °C has been adopted.

Water content of a soil in its *in situ* (natural) state is known as the *natural water content.* Values of typical water contents of natural soils are given in Table 3.3.

Laboratory determination of water contents should be carried out according to the British Standard 1377 (1967, Test 1).

It must be emphasized that, although this test is a very simple one, serious errors are frequently made, primarily owing to improper selection of representative samples. A soil sample, whether of undisturbed or disturbed type, is frequently non-homogeneous and may contain certain inclusions; a sample taken for the water content determination should contain sufficient material for other anticipated tests, e.g. determination of consistency limits or particle-size distribution. The whole of the soil sample should then be thoroughly and quickly mixed and subdivided into quantities appropriate to the anticipated tests; failure to observe this requirement may lead to erroneous determination of the consistency of the soil, because the determined water content may refer to a soil which is different from the one for which the consistency limits were obtained.

3.6.2. NATURAL DENSITY OF SOILS

The ratio of the mass of a soil to its volume (in its natural state) is known as the *natural density* or *bulk density*. In laboratory work it is usually expressed in terms of g/ml while in design calculations kg/m^3 or t/m^3 units are often used, numerically g/ml = t/m^3 (in Imperial Units density is expressed in lb/ft^3; 1·0 g/ml = 62·4 lb/ft^3 = density of water):

$$\rho = \frac{M}{V} \tag{3.12a}$$

where M = mass of the soil sample in g
V = volume of the soil sample in ml

The S.I. system of units differentiates very clearly between the mass and the gravitational force. Density of a given material depends on the mass and on the concentration of the constituent particles and therefore is one of its basic properties; it is independent of the location of the material, i.e. it is the same on the earth as it would be on the moon. On the other hand, the gravitational force per unit volume of a given material (commonly referred to as the unit weight or specific weight) is equal to the product of its density and the gravitational acceleration 'g'; it is a function of 'g' and therefore it is not a basic property of the material.

Therefore when referring to the physical properties of the materials the term density should be used as defined by Equation (3.12a) and denoted by ρ. When, however, the effects on the gravitational force are to be considered the term 'unit weight' should be used and denoted by γ. By definition:

$$\gamma = \rho g \tag{3.12b}$$

where γ = unit weight in kN/m^3
ρ = density in t/m^3 ($t/m^3 = g/ml = kg/l$)
g = gravitational acceleration in m/s^2 ($g \approx 9{\cdot}81\ m/s^2$)

The density of soils depends on the density of the solid particles, on the porosity, and on the water content. Typical values of densities of natural soils are given in Table 3.3.

While the determination of the mass of a soil sample, apart from preservation of it natural water content, presents no difficulties the measurement of its *in situ* volume may present problems. In the case of the fine-grained soils (particularly of the cohesive type) it is usually possible to obtain undisturbed, regular or irregular, samples and hence their volume can be either computed or measured. It is, however, seldom possible to obtain undisturbed samples of soils containing grains of gravel size or larger and therefore in such cases (particularly in road construction) the volume of the hole from which a sample of a known mass has been removed is measured.

Determination of densities should be carried out in accordance with the British Standard 1377 (1967, Tests 14A to F).

If it is anticipated that in addition to determination of the density of a soil (contained in a single undisturbed sample) other physical properties are to be determined, then the density should be determined first, before the sample is broken down and thoroughly mixed prior to the other tests, e.g. water content or consistency determination. Results of proper tests should compare with the figures given in Table 3.3.

3.6.3. DENSITY OF SOLID PHASE

Density of the solid phase is the ratio of the mass of the dry solid particles to their volume. It is usually expressed in terms of g/ml or as specific gravity G_s (i.e. as a dimensionless ratio of the unit weight of solids to the unit weight of water, $G_s = g\rho_s/g\rho_w = \rho_s/\rho_w$):

$$\rho_s = \frac{M_s}{V_s} \tag{3.13a}$$

where M_s = mass of the dry solid particles in g
V_s = volume of the same solid particles (soil skeleton) in ml

The specific gravity of solid particles depends on the mineralogical composition of the soil and variės between 2·40 and 3·20; average specific gravities of individual types of particles in soils are as follows:

(1) kaolinite 2·4
(2) quartz 2·65
(3) limestone 2·72
(4) dolomite 2·80–2·95
(5) mica 2·70–3·20

For quartz sands and loess and for other soils which do not contain organic matter or heavy minerals it can be assumed that on average

$$G_s = 2{\cdot}65$$

Typical values of specific gravities of solid particles to be found in the majority of the soils are shown in Table 3.3.

Determination of density (or specific gravity) of the solid phase should be carried out in accordance with the British Standard 1377 (1967, Tests 6A and B).

The unit weight of solid particles is, as in the case of the unit weight of soils, obtained by multiplying the density ρ_s by the gravitational acceleration 'g':

$$\gamma_s = g\rho_s = gG_s\rho_w \tag{3.13b}$$

For practical purposes $\rho_w = 1{\cdot}0$ g/ml = $1{\cdot}0$ t/m^3 and $g = 9{\cdot}81$ m/s^2 and therefore

$$\gamma_s = 9{\cdot}81\ G_s\ \text{kN/m}^3 \tag{3.13c}$$

3.7. Characteristics Defining Porosity of Soils

Having determined the basic physical properties, i.e. the water content w, the natural density ρ, and the density of the solid phase ρ_s (or the specific gravity G_s) of a given soil, it is now possible to determine other characteristics relating to the arrangement of the grains and particles, i.e. defining the porosity of the soil.

These characteristics include dry density of the soil, its porosity, and voids ratio. They are an indirect measure of the compressibility of the soil under loading and have therefore a direct and extensive application in geotechnical engineering.

3.7.1. DRY DENSITY OF SOILS

The dry density of a soil is defined as the dry mass of the grains and particles per unit volume of the soil; it is expressed in g/ml:

$$\rho_d = \frac{M_s}{V} \tag{3.14}$$

where M_s = dry mass of the solid particles in the sample in g
V = volume of the sample (before drying) in ml

The dry density is used in the evaluation of porosity and voids ratio. It is particularly frequently used in road construction for determination of the state of compaction of fill materials.

In the case of evaluation of ρ_d from the knowledge of water content w and natural density ρ, Equation (3.15) can be used which has been derived from the relationships shown in Figure 3.8:

$$\rho = \frac{M}{V} = \frac{M_s + M_w}{V} = \frac{M_s + w\,M_s/100}{V} = \rho_d + \frac{w}{100}\rho_d$$

and, on rearrangement,

$$\rho_d = \frac{\rho}{100 + w} \times 100 \tag{3.15a}$$

By definition the dry unit weight γ_d is equal to

$$\gamma_d = g\rho_d = \frac{g\rho}{100 + w} \times 100 \tag{3.15b}$$

and for practical purposes is expressed in kN/m^3.

3.7.2. POROSITY

Porosity n (dimensionless) is defined as a ratio of the volume of pores V_p to the total volume of the soil V (solid phase + pores); therefore in mathematical terms

$$n = \frac{V_p}{V} \tag{3.16}$$

Table 3.3. Typical values of specific gravities G_s of solids, natural water contents w in %, and bulk densities ρ in g/ml (after Polish Standard PN-59/B-03020, 1959)

Type of soil			State of saturation (Table 3.7)	Specific gravity of solids G_s	Water content w / Bulk density ρ	State of compaction of cohesionless soils: Dense	Medium dense	Loose
						Density index I_D: 1·0 – 0·67	0·67 – 0·33	0·33 – 0
cohesionless	inorganic	gravels, tills, hoggins, residual rock debris (granite), etc.	damp	2·65	w/ρ	3/1·85	4/1·75	5/1·70
			moist		w/ρ	10/2·00	12/1·90	15/1·85
			wet		w/ρ	14/2·10	18/2·05	23/2·00
		quartzitic sands: coarse and medium (silt content < 10%, clay content < 2%)	damp	2·65	w/ρ	4/1·80	5/1·70	6/1·65
			moist		w/ρ	12/1·90	14/1·85	16/1·80
			wet		w/ρ	18/2·05	22/2·00	25/1·95
		sands: fine and silty (silt content < 30%, clay content < 2%)	damp	2·65	w/ρ	5/1·70	6/1·65	7/1·60
			moist		w/ρ	14/1·85	16/1·75	19/1·70
			wet		w/ρ	22/2·00	24/1·90	28/1·85
	organic	sands: organic	damp	2·64	w/ρ	5/1·60	6/1·55	7/1·50
			moist		w/ρ	16/1·75	18/1·70	21/1·65
			wet		w/ρ	24/1·90	28/1·85	30/1·75

Table 3.3 (contd.)

		Type of soil	Specific gravity of solids G_s	Water content w / Bulk density ρ	Consistency of cohesive soils: Hard or very stiff	Stiff	Firm	Soft to very soft
					Consistency index I_c: w_s 1·0	0·75	0·50	0
cohesive	inorganic	slightly clayey sand	2·65	w/ρ	10*/2·20	13/2·15	16/2·10	19/2·05
cohesive	inorganic	sandy silt	2·66	w/ρ	14*/2·15	17/2·10	19/2·05	22/2·00
cohesive	inorganic	silt	2·67	w/ρ	18*/2·10	21/2·05	23/2·00	26/1·95
cohesive	inorganic	clayey sand	2·67	w/ρ	9/2·25	12/2·20	17/2·10	24/2·00
cohesive	inorganic	clayey sandy silt	2·67	w/ρ	13/2·20	16/2·15	21/2·05	27/1·95
cohesive	inorganic	clayey silt	2·68	w/ρ	17/2·15	20/2·10	25/2·00	32/1·90
cohesive	inorganic	sand–clay	2·68	w/ρ	11/2·25	14/2·15	20/2·05	30/1·95
cohesive	inorganic	sand–silt–clay	2·69	w/ρ	15/2·20	18/2·10	24/2·00	35/1·90
cohesive	inorganic	silt–clay	2·71	w/ρ	18/2·15	22/2·00	28/1·90	42/1·80
cohesive	inorganic	sandy clay	2·70	w/ρ	14/2·20	18/2·10	25/1·95	40/1·80
cohesive	inorganic	clay	2·72	w/ρ	17/2·15	22/2·00	30/1·85	45/1·75
cohesive	inorganic	silty clay	2·75	w/ρ	20/2·05	25/1·90	33/1·80	50/1·70
cohesive	organic	silts with traces of organic matter	2·30	w/ρ	20–40/2·00–1·80			
cohesive	organic	organic alluvial muds	2·15 to 2·60	w/ρ	20–150/1·90–1·30			
cohesive	organic	peats	1·50 to 2·15	w/ρ	25–400/1·80–1·00			

* For slightly cohesive soils having smaller water contents than tabulated natural densities are lower and depend on the degree of saturation–see nomograms in Chapter 7.

Because of the difficulties in a direct determination of the volume of the solid phase and pores an indirect method, based on the relationships shown in Figure 3.10, is used:

$$M_s = V_s \rho_s$$

$$\rho_d = \frac{M_s}{V} = \frac{V_s \rho_s}{V} = (1 - n)\, G_s \rho_w$$

Rearranging the last expression porosity n can be expressed in terms of G_s and ρ_d

$$n = \frac{G_s \rho_w - \rho_d}{G_s \rho_w} \tag{3.17}$$

where $G_s\rho_w$ = density of soil particles in g/ml
ρ_d = dry density of soil in g/ml

Assuming that the soil consists of spherical grains of the same diameter d, the following results would be obtained.

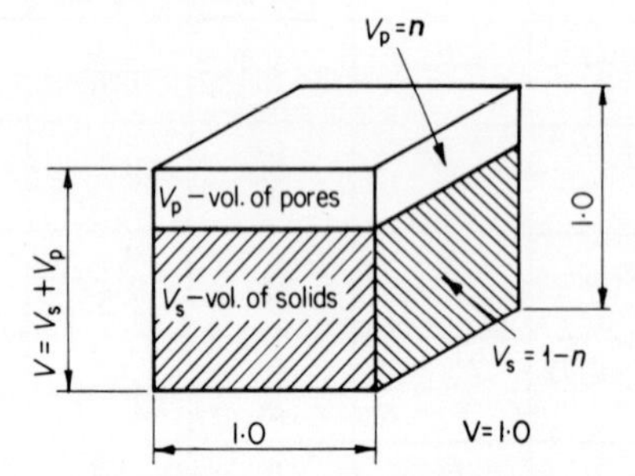

Figure 3.10. Volume of soil, solid phase, and pores.

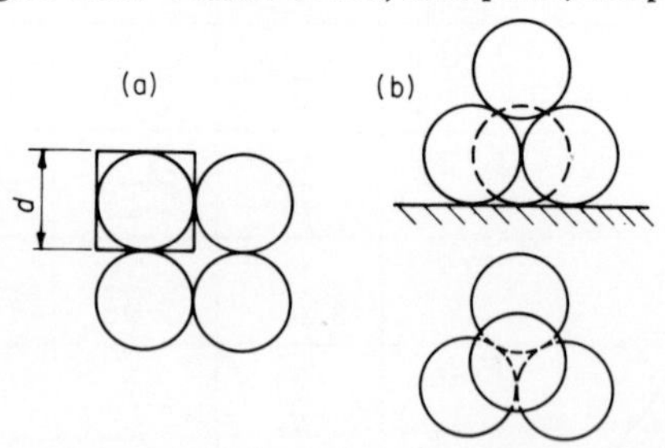

Figure 3.11. Arrangement of spherical particles.

Maximum porosity for 'sphere over sphere' arrangement (Figure 3.11(a)) would be

$$n_{max} = \left(d^3 - \frac{\pi d^3}{6}\right) \Big/ d^3 = 0{\cdot}476$$

and minimum porosity for 'sphere resting on three spheres' arrangement (Figure 3.11(b))

$$n_{min} = 0{\cdot}258$$

Porosities of uniformly graded sands and gravels vary between these limits; well-graded sands can have lower porosities than the above figures.

Cohesive soils have considerably higher porosities because clay particles arranged in honeycomb or flocculent structures result in much more porous systems; the pores in these soils are to a large extent filled with the adsorbed water, strongly attracted to the surface of the particles.

3.7.3. VOIDS RATIO

Voids ratio e (dimensionless) is defined as the ratio of the volume of voids (pores) to the volume of the solid phase (solids); it is determined from the following expression:

$$e = \frac{V_p}{V_s} = \frac{n}{1-n} = \frac{G_s \rho_w - \rho_d}{\rho_d} \tag{3.18}$$

For sands and gravels the voids ratio varies between 0·3 and 1·0; for cohesive soils these values are, of course, considerably higher.

Example. An undisturbed sample of sand had a volume of 1·0 litres and at its natural water content its mass was 1·845 kg. After drying, its mass was found to be 1·674 kg. Determine its voids ratio.

Firstly its water content is determined from Equation (3.11):

$$w = \frac{1{\cdot}845 - 1{\cdot}674}{1{\cdot}674} \times 100 \approx 10{\cdot}2\%$$

and its density ρ according to Equation (3.12a) is

$$\rho = \frac{1845}{1000} = 1{\cdot}85 \text{ g/ml}$$

It is now possible to determine its dry density ρ_d from Equation (3.15a):

$$\rho_d = \frac{1{\cdot}85}{100 + 10{\cdot}2} \times 100 \approx 1{\cdot}67 \text{ g/ml}$$

or, according to Equation (3.14),

$$\rho_d = \frac{1674}{1000} \approx 1{\cdot}67 \text{ g/ml}$$

To evaluate its porosity or voids ratio it is now necessary to know the density of the solid phase; for sand, with sufficient degree of accuracy, it can be taken that $G_s = 2{\cdot}65$ and that $\rho_w = 1{\cdot}0$ g/ml (see Table 3.3). Using Equation (3.17) its porosity can be obtained:

$$n = \frac{2{\cdot}65 \times 1{\cdot}0 - 1{\cdot}67}{2{\cdot}65 \times 1{\cdot}0} \approx 0{\cdot}37$$

and, from Equation (3.18), its voids ratio

$$e = \frac{2{\cdot}65 \times 1{\cdot}0 - 1{\cdot}67}{1{\cdot}67} = 0{\cdot}59$$

Evaluation of all the above characteristics from the knowledge of w and ρ can very speedily be carried out with the help of the nomograms in Chapter 7; the degree of saturation and saturation water content can also be read off from these nomograms.

3.8. Density Index and States of Compaction of Granular Soils

In the determination of safe bearing stresses for granular soils (sands and gravels) one has to consider their state of compaction (density of packing) which is defined by the 'density index', also known as 'relative density'.

3.8.1. DENSITY INDEX

The density index of cohesionless soils I_D (dimensionless) is defined as the ratio of the natural density of packing to the possible maximum density of packing of the given soil.

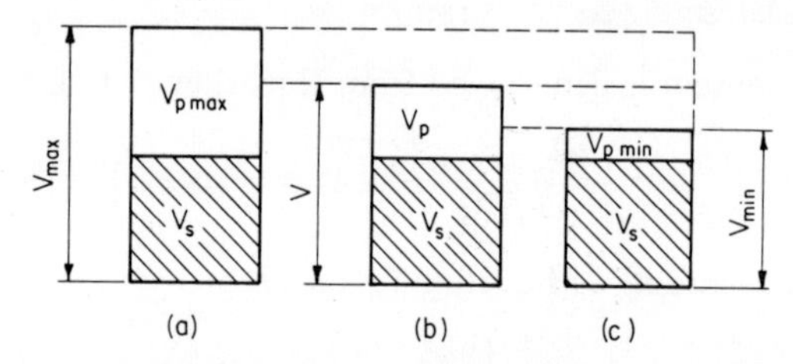

Figure 3.12. Changes in volume of voids in sand with increase in packing; (a) volume at loosest packing; (b) volume at natural packing; (c) volume at densest packing.

The density index is computed in accordance with details given in Figure 3.12 from the following expression:

$$I_D = \frac{V_{max} - V}{V_{max} - V_{min}} = \frac{V_{p\,max} - V_p}{V_{p\,max} - V_{p\,min}} = \frac{\dfrac{V_{p\,max}}{V_s} - \dfrac{V_p}{V_s}}{\dfrac{V_{p\,max}}{V_s} - \dfrac{V_{p\,min}}{V_s}} = \frac{e_{max} - e}{e_{max} - e_{min}} \tag{3.19}$$

where e_{max} = maximum voids ratio obtained at loosest packing
e_{min} = minimum voids ratio obtained at densest packing
e = natural voids ratio

The minimum voids ratio e_{min} or the maximum dry density $\rho_{d\ max}$ of a granular soil can be determined by compacting it in a standard compaction mould using a Kango hammer. The mould is filled in three equal layers and each is compacted for 1½ minutes. The minimum voids ratio is evaluated from the determined $\rho_{d\ max}$ and the known G_s from Equation (3.18).

There are two generally accepted methods for the determination of the maximum voids ratio e_{max} or the minimum dry density $\rho_{d\ min}$ of granular soils: the 'funnel method' and 'cylinder method' (Akroyd, 1957).

In the funnel method a given soil is slowly poured through a funnel into a standard compaction mould, with the funnel outlet about 5 mm from the surface of the sample in the mould, and from the known volume and measured mass of the soil its minimum dry density $\rho_{d\ min}$ is determined.

In the cylinder method a known mass of the given soil is poured into a standard measuring cylinder. The palm of the hand is placed over the top of the cylinder, which is shaken a few times and turned upside down. It is then quickly turned the right way up and the volume occupied by the sample is read off. The test is repeated several times and the maximum reading is recorded from which $\rho_{d\ min}$ is determined.

In both cases the maximum voids ratio is evaluated from the determined $\rho_{d\ min}$ and the known G_s from Equation (3.18).

The natural voids ratio e is determined from the knowledge of w, ρ, and G_s from Equations (3.15) and (3.18).

3.8.2. STATES OF COMPACTION

Mechanical properties of granular soils depend on their state of compaction which can be defined by the density index I_D. A suggested simple descriptive classification covering three ranges of I_D is given below (after the Polish Standard PN-54/B-02480, 1954).

Density index	Descriptive term
$I_D \leqslant 0{\cdot}33$	loose
$0{\cdot}33 < I_D \leqslant 0{\cdot}67$	medium dense
$0{\cdot}67 < I_D \leqslant 1{\cdot}0$	dense

A more comprehensive classification has been developed for the interpretation of the standard penetration test results (see Section 6.3.1).

Density index	Descriptive term
$I_D \leqslant 0{\cdot}15$	very loose
$0{\cdot}15 < I_D \leqslant 0{\cdot}35$	loose
$0{\cdot}35 < I_D \leqslant 0{\cdot}65$	medium dense
$0{\cdot}65 < I_D \leqslant 0{\cdot}85$	dense
$0{\cdot}85 < I_D \leqslant 1{\cdot}0$	very dense

Example. In its natural state a sample of sand had a volume of 1·0 litres, on drying and gently pouring into a measuring cylinder its volume was 1·22 litres, and on compaction by vibration 0·92 litres. Determine its state of compaction.

The density index I_D is determined from Equation (3.19):

$$I_D = \frac{V_{max} - V}{V_{max} - V_{min}} = \frac{1{\cdot}22 - 1{\cdot}0}{1{\cdot}22 - 0{\cdot}92} = 0{\cdot}71$$

Therefore the sand is in a *dense* state.

3.9. Limits of Consistency, Consistency Index, and Consistency of Cohesive Soils

The mechanical properties of cohesive soils mainly depend on their water content, on the mineralogical composition of the clay fraction, and on the type of the adsorbed cations (Chapters 2 and 5). Since the last two factors also govern the consistency limits (plastic and liquid limit), then the description of these soils in terms of their state of consistency, which directly depends on the water content and the consistency limits, will also describe them according to their basic mechanical properties.

3.9.1. LIMITS OF CONSISTENCY OF COHESIVE SOILS

We can distinguish between three states of consistency of cohesive soils: liquid, plastic, and solid.

A soil of *liquid consistency* behaves like a fluid and its resistance to shearing is almost negligible.

A soil of *plastic consistency* can be readily moulded without formation of cracks and will retain its new shape.

A soil of *solid consistency* will only deform under high stresses and the deformations are accompanied by formation of cracks.

The boundary between the liquid and plastic consistency is known as the *liquid limit* and that between the plastic and solid consistency the *plastic limit.*

Apart from the above limits we also consider the *shrinkage limit* which a given soil of solid consistency reaches on drying, when it ceases to decrease in volume.

The *liquid limit* w_l is determined from an empirical test; it is the water content (in percentages) of a soil paste (soil–water mixture) which, on being placed in the cup of Casagrande's apparatus and grooved with a special tool, will cause the groove to close on the twenty-fifth blow of the cup against the base of the apparatus.

Because of its empirical character the liquid limit must be determined strictly in accordance with the British Standard 1377 (1967, Tests 2A or B).

The *plastic limit* w_p is defined as the water content (in percentages) of the soil when, on repeated rolling of a small ball of the soil into a thread, the soil crumbles when the thread is 3 mm ($\frac{1}{8}$ in) in diameter.

Again this limit must be determined strictly in accordance with the British Standard 1377 (1967, Test 3).

In the case when a soil contains a large proportion of grains greater than 0·5 mm then the consistency limits are determined using only that soil which passes the No. 36 (0·42 mm mesh) B.S. test sieve; the results must quote the proportion (by dry weight) of the fine material passing the No. 36 sieve.

The *shrinkage limit* w_s is defined as the water content (in percentages) of the soil at the stage when on drying it ceases to decrease in volume. In practice

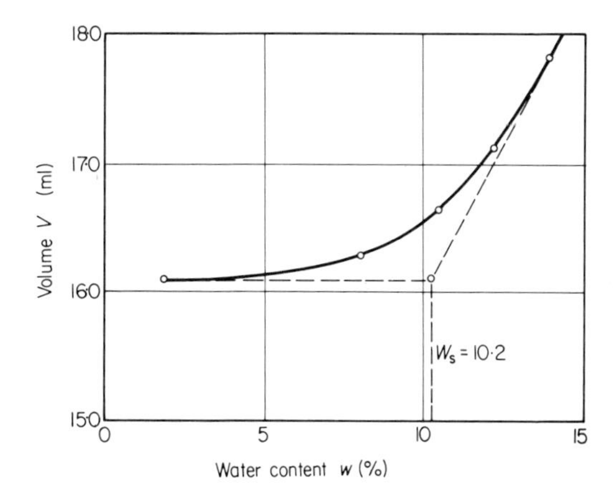

Figure 3.13. Determination of shrinkage limit.

the shrinkage limit is taken as that moisture content which corresponds to the point of intersection of the tangents to the experimental volume–water content curve (Figure 3.13).

For foundation purposes the shrinkage limit is determined when it is necessary to estimate the effects of water content changes that may take place after construction, and also in the case when the natural water content of a given soil is found to be below the plastic limit and it is necessary to establish whether the soil is of the hard or the very hard consistency.

The test is carried out using an undisturbed sample of the soil of natural water content and between 20 and 30 ml in volume (without gravel grains). Initially the sample is allowed to dry in the air until it begins to change to lighter colour tones; drying is then continued in an oven in which the temperature is gradually increased until the maximum of 105–110 °C is reached. During the drying process measurements of the mass and volume (by immersion in mercury) of the sample are taken at certain time intervals until both of them become constant. The dry mass M_s of the sample is then used in evaluation of the water content at the various stages of drying.

On the basis of the above results the volume–water content curve is plotted from which the shrinkage limit w_s is determined (Figure 3.13).

3.9.2. LIQUIDITY AND CONSISTENCY INDEXES

Liquidity index I_l (dimensionless) is defined as the ratio of the difference of the natural water content and plastic limit to the difference of the liquid limit and plastic limit:

$$I_l = \frac{w - w_p}{w_l - w_p} \tag{3.20}$$

where w = natural water content in %
w_p = plastic limit in %
w_l = liquid limit in %

Consistency index I_c (dimensionless) is defined as the ratio of the difference of the liquid limit and natural water content to the difference of the liquid limit and plastic limit:

$$I_c = \frac{w_l - w}{w_l - w_p} \tag{3.21}$$

The two indices are related by Equation (3.22):

$$I_c = 1 - \frac{w - w_p}{w_l - w_p} = 1 - I_l \tag{3.22}$$

3.9.3. CONSISTENCY OF COHESIVE SOILS

The consistencies of cohesive soils can be classified either on the basis of their physical properties such as natural water content and consistency limits (using the consistency index I_c) or on the basis of mechanical properties such as the undrained shear strength (Table 3.4).

Table 3.4. Consistencies of cohesive soils

States of consistency and descriptive terms	Values of I_c and w	Undrained* (immediate) shear strength (kN/m^2)
very hard	$I_c > 1{\cdot}0$ and $w \leqslant w_s$	
hard or very stiff	$I_c > 1{\cdot}0$ $w_s < w \leqslant w_p$	greater than 144†
stiff	$1{\cdot}00 > I_c \geqslant 0{\cdot}75$	72 to 144
firm	$0{\cdot}75 > I_c \geqslant 0{\cdot}50$	36 to 72
soft	$\{0{\cdot}50 > I_c \geqslant 0{\cdot}00$	18 to 36
very soft		less than 18
liquid	$I_c < 0{\cdot}0$ or $w > w_l$	

* Usually unconfined compression test.
† For approximate conversion to lbf/ft^2 multiply by 21·0 and to kgf/cm^2 by 0·01.

The classification recommended by the authors (second column in Table 3.4) is illustrated in Figure 3.14.

The classification based on the undrained shear strength may give different results in the case of sensitive soils. The sensitive soils in the remoulded state (at natural water content) usually appear softer because the remoulding processes destroy the effects of cementation or of diagenetic bonds that may exist between the particles. However, the influence of any interparticle bonds on the mechanical properties of the soils, other than the true cohesion, cannot always be considered as permanent (Sections 2.6.1 and 5.4.3).

It is, therefore, considered that a classification which eliminates them, but takes into account the basic factors (the natural water content, the mineralogical composition of particles, and the type of adsorbed cations) that govern the mechanical properties of cohesive soils, is the most objective and fundamental in its approach.

The classification of consistencies of cohesive soils suggested by the British Standard Code of Practice CP 2001 (1957) is based on the undrained (unconfined) shear strength of soils in their natural (undisturbed) state (third column in

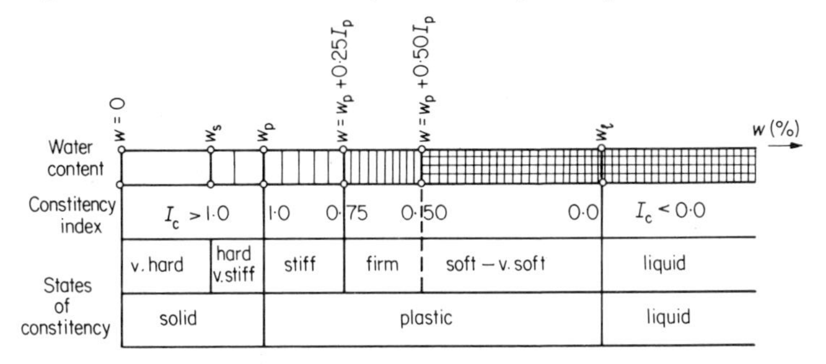

Figure 3.14. Consistency of cohesive soils with relation to their natural water content and consistency limits.

Table 3.4). In the case of insensitive soils of plastic consistency the two classifications give comparable descriptions of consistencies. However, in the case of soils in the 'stiff' and 'very stiff' consistency ranges the undrained shear strength may be affected by sample disturbance or by presence of fissures and hence may lead to an incorrect classification.

3.9.4. PLASTICITY INDEX

Plasticity index I_p is defined as the difference between the liquid and plastic limit and is expressed in percentages:

$$I_p = w_l - w_p \tag{3.23}$$

The plasticity index indicates the quantity of water in percentages (of the dry mass of soil) that a given soil absorbs in changing from the semi-solid to liquid consistency. Highly active sodium bentonites, which readily absorb water, have plasticity indices over 200% while slightly active loesses (quartzitic silts) have plasticity indices between 5 and 10%. Soils possessing low plasticity indices can very easily liquefy, even at very low water contents.

The ratio of the plasticity index I_p to the clay fraction J (in percentages) in a given soil has been defined by Skempton (1953) as its activity

$$A = \frac{I_p}{J} \tag{3.24}$$

where A = 'activity' or 'colloidal activity' of clays, dimensionless
J = clay fraction (particles smaller than 2 μm) expressed as percentage of the dry mass of soil

For a large proportion of the soils in the British Isles the activity varies between 0·75 and 1·25 and can be taken as approximately 1·0; the exceptions

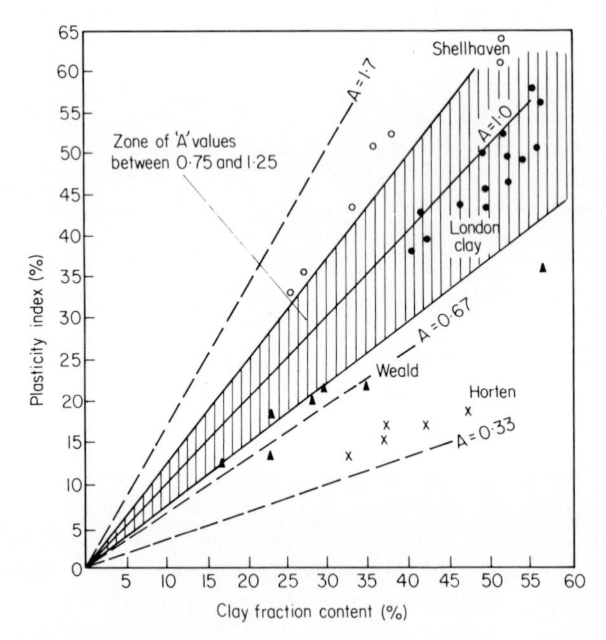

Figure 3.15. Plasticity index and clay fraction content relationship for the majority of soils encountered in the British Isles.

are kaolins, late glacial and post-glacial lacustrine clays and loesses for which A varies between 0·4 and 0·75, and organic estuarine clays for which A is greater than 1·25 (Figure 3.15).

Taking the above into consideration it can be assumed that for most of the cohesive soils $I_p = J$ and hence these two quantities can be used for their approximate classification as shown in Table 3.5 (after the Polish Standard PN-54/B-02480 (1954)).

Table 3.5. Classification of cohesive soils

Description of cohesiveness of soil	Clay fraction ($<2\ \mu m$) content J (%)	Plasticity index I_p (%)
cohesionless	0– 2	<1
slightly cohesive	2– 10	1–10
medium cohesive	10– 20	10–20
cohesive	20– 30	20–30
very cohesive	30–100	>30

In the case of a discrepancy in the description of a soil by the plasticity index (third column) and the clay fraction content (second column), the colloidal activity should be quoted after the description, e.g. for a soil with $J = 25\%$ and $I_p = 38\%$ the following should be stated: heavy cohesive silt–clay ($J = 25\%$, $A = 1{\cdot}5$).

3.9.5. CLASSIFICATION OF COHESIVE SOILS BY MEANS OF PLASTICITY CHART

According to the British Standard Code of Practice CP 2001 (1957, Appendix E), the most widely used soil classification system for roads and airfields is the one developed by A. Casagrande (1948). In this classification system the soil type is designated by two capital letters which indicate the main soil types and certain of their characteristics (Table 3.6).

Table 3.6

	Soil type/descriptive characteristic		Prefix
Main soil type	coarse-grained	gravel	G
		sand	S
	fine-grained	silt	M
		clay	C
	organic	organic silts and clays	O
		peat	Pt
Descriptive characteristic	coarse-grained	well graded	W
		poorly graded	P
		uniformly graded	U
		containing excess of fines	F
	fine-grained	low plasticity $w_l < 35\%$	L
		medium plasticity $35\% \leqslant w_l < 50\%$	I
		high plasticity $w_l \geqslant 50\%$	H

In order to assign the appropriate letters to the fine-grained soils Casagrande's plasticity chart is used (Figure 3.16). The plasticity index of a soil is plotted against its liquid limit, and the classification letters are determined from the position of the point on the chart. The line 'A' is taken as a boundary between the organic and inorganic soils, the latter lying above the line. Fine-grained soils are subdivided according to their plasticity as shown in Table 3.6 and in Figure 3.16. For soils with $w_1 < 30\%$ the choice of classification letters is also based on visual inspection and particle-size analysis.

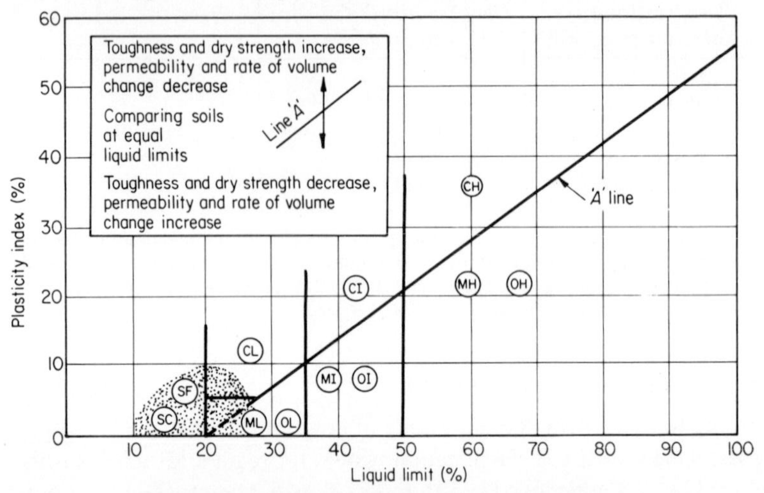

Figure 3.16. Plasticity chart for soil classification.

Because of its simplicity the plasticity chart is frequently used in the British Isles for general classification of cohesive soils in all types of site investigation work, i.e. for foundation design as well as for roads and airfields.

3.9.6. DETERMINATION OF CONSISTENCY INDEX BY THREAD-ROLLING METHOD

Using Equation (3.22) and assuming that during one thread-rolling operation the soil loses approximately 1·25% of its water content (Figure 3.17), the consistency index can be determined by the thread-rolling method according to Equation (3.25) (Wiłun, 1951):

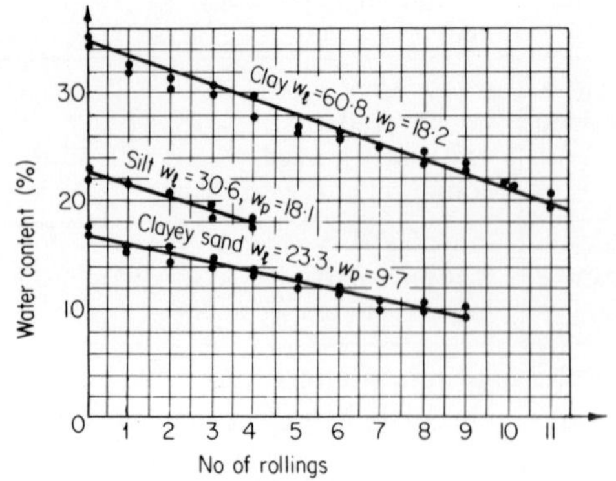

Figure 3.17. Variation of water content with the number of thread-rolling operations.

$$I_c = 1 - \frac{w - w_P}{I_p} = 1 - \frac{1{\cdot}25N}{AJ} \tag{3.25}$$

In the fractional part of the above equation the numerator indicates the loss in water content during N thread-rolling operations (from natural water content to plastic limit) while the denominator is the plasticity index as given by Equation (3.24).

The above method enables one to determine the consistency index in the field as well as in the laboratory (the Polish Standard PN-55/B-04482, 1955); detailed procedure is described in Section 3.12.2.

3.10. Effects of Water Content on Density and Unit Weight of Soils

Depending on the degree of saturation, soil densities and unit weights can be evaluated from different expressions.

If the pores are completely filled with water (saturated) but the soil is situated above the ground water level then its density is

$$\rho_{sat} = (1 - n)\, G_s \rho_w + n\rho_w \tag{3.26a}$$

where ρ_{sat} = density of fully saturated soil in g/ml
$(1 - n)G_s\rho_w$ = mass of solids per unit volume of soil in g/ml
$n\rho_w$ = mass of water filling the pores per unit volume of soil in g/ml

Again, by definition, the unit weight of saturated soil is equal to

$$\gamma_{sat} = g\,\rho_{sat} = (1 - n)\, G_s\gamma_w + n\gamma_w \tag{3.26b}$$

and for practical purposes it is expressed in kN/m^3.

If a soil is situated below the ground water level, then its pores are completely filled with water but its unit effective weight* is considerably smaller than γ_{sat} because the water is not supported by the soil skeleton but it exerts an uplift on it (according to Archimedes' principle):

$$\gamma_{sub} = (1 - n)G_s\gamma_w - (1 - n)\gamma_w = (1 - n)(G_s - 1)\gamma_w \tag{3.27}$$

where γ_{sub} = submerged unit weight of soil skeleton in kN/m^3
$(1 - n)G_s\gamma_w$ = weight of solids per unit volume of soil in kN/m^3
$(1 - n)\gamma_w$ = weight of water displaced by the solids per unit volume of soil (uplift) in kN/m^3

The difference between γ_{sat} and γ_{sub} can be evaluated by subtracting Equation (3.27) from (3.26b); assuming ρ_w = 1·0 g/ml, and hence $\gamma_w = g\rho_w$ = 9·8 kN/m^3, we obtain

$$\gamma_{sat} - \gamma_{sub} = \gamma_w \approx 9{\cdot}8 \text{ kN/m}^3 \tag{3.28}$$

* In Archimedes' principle one is considering the forces acting on the soil skeleton of constant concentration of mass and therefore one can only consider unit weight.

With sufficient accuracy for practical purposes the above equation enables us to assume that for cohesive soils

$$\gamma_{sub} = (\gamma - 9.8)\ \text{kN/m}^3 \tag{3.29}$$

In this way unnecessary calculations are eliminated while the introduced errors are small because most of the cohesive soils, with the exception of soils in a solid state ($w < w_s$), are practically fully saturated.

3.11. Degrees of Saturation of Soils

In order to establish the state of saturation of a given soil and to determine to what extent the pores are filled with water it is necessary to evaluate the saturation water content and the degree of saturation.

3.11.1. SATURATION WATER CONTENT

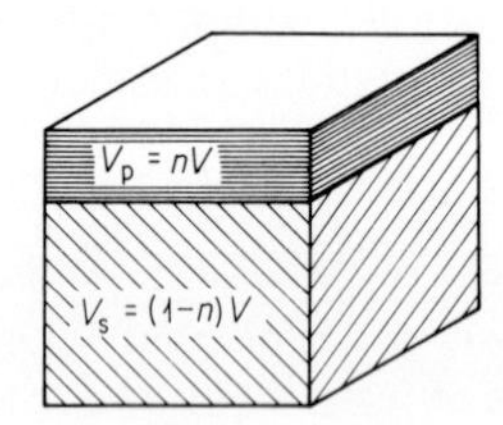

$M_w = nV\rho_w$, $M_s = (1-n)VG_s\rho_w$

Figure 3.18. Model of saturated soil.

A soil is fully saturated when its pores are completely filled with water (Figure 3.18). Saturation water content w_{sat}, in percentages, is evaluated from the following expression:

$$w_{sat} = \frac{M_w}{M_s}100 = \frac{nV\rho_w}{(1-n)VG_s\rho_w} \times 100 = \frac{e}{G_s}100 \tag{3.30}$$

where e = voids ratio

G_s = specific gravity of solids (dimensionless)

ρ_w = density of water = 1·0 g/ml

3.11.2. DEGREE OF SATURATION OF SOILS

The degree of saturation S_r defines to what extent the pores are filled with water (S_r = 0 for dry and 1·0 for fully saturated soil). It is evaluated from the following expression:

$$S_r = \frac{V_w}{V_p} = \frac{V_w\rho_w}{V_p\rho_w} = \frac{M_w}{(w_{sat}/100) \times M_s} = \frac{(M_w/M_s)100}{w_{sat}} = \frac{w}{w_{sat}} = \frac{wG_s}{100e} \tag{3.31}$$

where V_w = volume of water partly filling the pores in ml
V_p = volume of pores (voids) in ml
M_w = mass of water in the soil in g
M_s = mass of dry solids in the soil in g
w = natural water content in %
w_{sat} = saturation water content in %
G_s and e = as in Equation (3.30)

3.11.3. STATES OF SATURATION OF COHESIONLESS SOILS

The states of saturation of cohesionless soils are classified on the basis of the degree of saturation S_r as shown in Table 3.7.

Table 3.7. States of saturation of cohesionless soils

Degree of saturation	State of saturation of soil
$0 < S_r \leqslant 0{\cdot}4$	dry to damp
$0{\cdot}4 < S_r \leqslant 0{\cdot}8$	moist
$0{\cdot}8 < S_r \leqslant 1{\cdot}0$	wet

For fully saturated soils S_r is equal to 1·0. In all other cases air (water vapour) is present in the pores.

The presence of free air or gases (undissolved) has a considerable influence on a number of its properties; e.g. the angle of internal friction of a soil with $S_r < 0{\cdot}9$ is greater than that of the same soil but with $S_r = 1{\cdot}0$; also the process of settlement of the former is much more rapid than that of the latter.

Example. Determine the state of saturation of the sand described in the example in Section 3.7.3. The following data are taken from the previous example:

$$w = 10{\cdot}2\%, \quad e = 0{\cdot}59, \quad G_s = 2{\cdot}65$$

The saturation water content is evaluated from Equation (3.30):

$$w_{sat} = \frac{0{\cdot}59}{2{\cdot}65} \times 100 = 22{\cdot}2\%$$

The degree of saturation from Equation (3.31) is

$$S_r = \frac{10{\cdot}2}{22{\cdot}2} = 0{\cdot}46$$

According to Table 3.7 the sand is moist.

We can also determine saturated and submerged unit weights of the same sand. From Equation (3.26a) ρ_{sat} is obtained:

$$\rho_{sat} = (1 - n)G_s\rho_w + n\rho_w = (1 - 0{\cdot}37)\,2{\cdot}65 + 0{\cdot}37 = 2{\cdot}04 \text{ g/ml}$$

The unit weight of saturated sand $\gamma_{sat} = g\rho_{sat} = 9{\cdot}81 \times 2{\cdot}04 = 20{\cdot}0$ kN/m^3. From Equation (3.28) γ_{sub} is evaluated:

$$\gamma_{sub} = \gamma_{sat} - 9{\cdot}8 = 10{\cdot}2 \text{ kN/m}^3$$

Using Equation (3.27)

$$\gamma_{sub} = (1 - n)(G_s - 1)\gamma_w = (1 - 0{\cdot}37)(2{\cdot}65 - 1{\cdot}0)\,9{\cdot}81 = 10{\cdot}2 \text{ kN/m}^3$$

The same quantities obtained from nomogram 7.2 differ only slightly from the above.

3.12. Sampling of Soils for Laboratory Testing and Their Description

The results of laboratory tests can only be considered as accurate and representative of the *in situ* conditions if the following hold.

(a) Correct sampling procedures have been adopted on site and the samples have not been disturbed or allowed to dry or to absorb water (Section 6.2.2).

(b) Laboratory tests have been carried out on representative and homogeneous samples, typical of the *in situ* conditions.

(c) The tests have been properly conducted.

The samples should be extracted, stored, and transported in such a manner that the natural water contents and densities are preserved and segregation of individual components does not take place.

It is necessary to check in the laboratory that on arrival from the site the samples are properly packed and sealed and have not undergone any changes such as loss of water content or segregation, etc. In addition, prior to the final selection of the samples for laboratory testing, it is necessary to classify the soils and to establish their consistencies. This detailed classification or description of the soils is done with the help of a rapid, but sufficiently accurate for practical purposes, macroscopic analysis. On superposition of the descriptions of the soils on geological sections of the site, samples can be selected for testing that are most relevant to the stability and safety of the proposed building or structure.

Disturbed samples of soils contained in plastic bags, boxes, or jars usually undergo segregation during transportation. The soil in such samples must, therefore, be thoroughly mixed prior to subsampling for individual tests. In order to ensure homogeneous samples this is done by quartering or by dividing into rectangles or using a special sample divider (riffle) as described in the British Standard 1377 (1967).

In the subdivision by quartering the large sample is spread evenly on a slab and is divided into four equal parts; two diagonally opposite quarters are then combined and the procedure repeated until the two quarters combine to sample of the required size. Alternatively, the evenly spread large sample is divided into a network of equal rectangles from which equal quantities of soil are taken to make up a sample of the required size.

Samples which have undergone any changes in the water content or which are not homogeneous should not be used for determining natural water contents, bulk densities, or for any other investigations.

Sufficient accuracy in the results of laboratory tests can only be achieved if the tests are carried out in accordance with the appropriate standards (where applicable, these have been quoted in the description of the individual tests); divergence from the standard methods may lead to considerable errors.

3.12.1. DESCRIPTION OF SOILS (MACROSCOPIC ANALYSIS)

The macroscopic analysis of soils involves their full description without the use of a microscope or other complicated pieces of apparatus.

The full description of a soil must include its following characteristics.

(a) Type of soil.
(b) Consistency or compaction.
(c) Structural characteristics.
(d) Colour.
(e) State of saturation (usually only in the case of cohesionless soils).
(f) Odour when present (for organic soils and peats).

Example. Firm dark-brown/grey intact sandy clay,
loose yellow fine sand (moist),
soft dark-brown/light-grey laminated silt–clay,
hard dark-grey weathered silty clay.

3.12.2. RECOGNITION OF TYPE AND CONSISTENCY OF COHESIVE SOILS

In the determination of the type and consistency of a cohesive soil the thread-rolling method is used. For this test a small lump of soil is taken from the centre of a larger piece and a ball of approximately 8 mm ($\frac{1}{3}$ in) in diameter is formed. The ball is placed on an open palm of one hand and with the palm cushion of the other (Figure 3.19) is rolled backwards and forwards along the axes of the open palm (at an approximate rate of two movements of the hand per second) under a slight pressure, until a thread of 3 mm ($\frac{1}{8}$in) in diameter is formed.

If the thread on reaching the diameter of 3 mm does not exhibit any damage, then it is tightly kneaded back into a ball and the thread-rolling process is repeated. The whole procedure is repeated until, on reaching the diameter of 3 mm, the thread begins to split or crumbles. The number of the thread-rolling operations leading to splitting of the thread is noted and the appearance of the cracks is carefully observed.

From the appearance of the thread the following conclusions with regards to the cohesiveness (active clay fraction content) of the soil can be drawn.

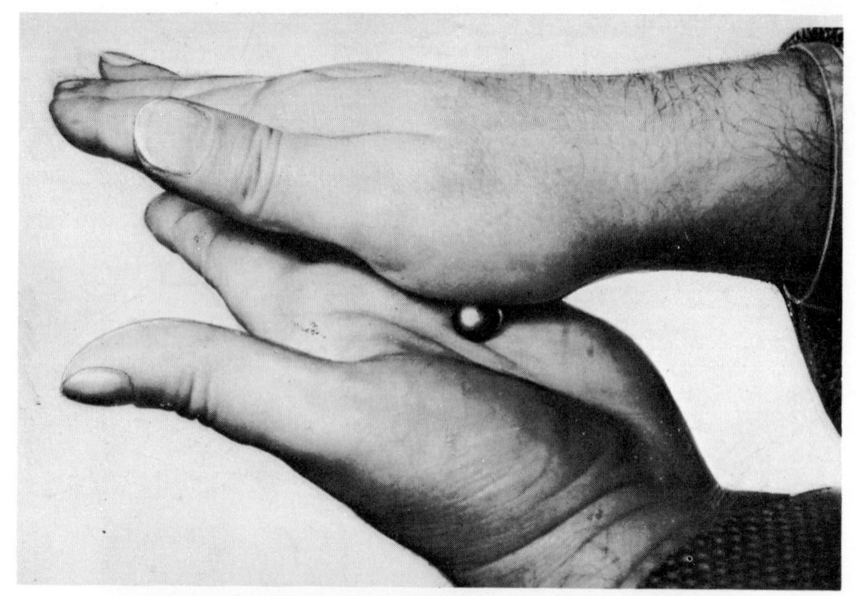

Figure 3.19. Beginning of the thread-rolling operation.

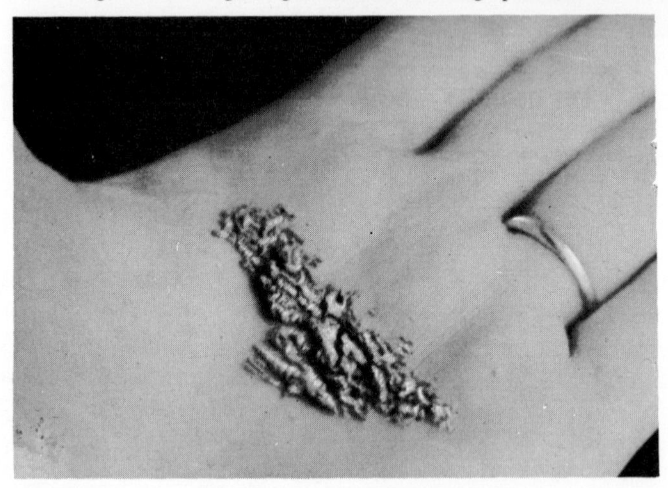

Figure 3.20. Crumbling of a thread of slightly cohesive soil ($J < 5\%$).

(a) If at the beginning of the rolling the thread flattens or crumbles (Figure 3.20) and cannot be rolled, then the soil is only slightly cohesive with active clay fraction content $J < 5\%$.

(b) If the thread crumbles, splitting longitudinally into layers (Figure 3.21) or if towards the end of rolling it has frayed ends, then the soil is slightly cohesive ($J = 5$ to 10%).

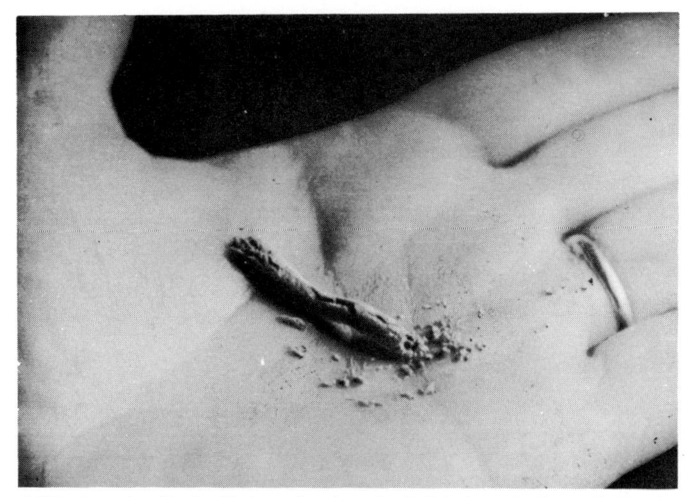

Figure 3.21. Longitudinal splitting of a thread of slightly cohesive soil (J = 5–10%).

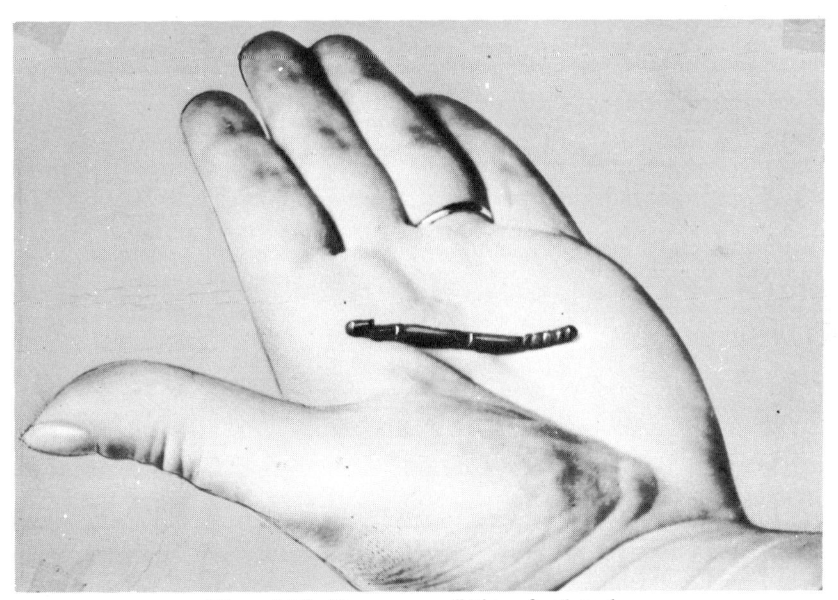

Figure 3.22. Transverse splitting of a thread.

(c) If, towards the end of rolling, the thread crumbles, splitting transversely (Figure 3.22), but all the time during rolling its surface has a dull, matt appearance, then the soil is medium cohesive (J = 10 to 20%).

(d) If, towards the end of rolling, the thread crumbles, splitting distinctly transversely, and its surface becomes shiny, although initially it was dull (the hand also becomes greasy in appearance), then the soil is cohesive (J = 20 to 30%).

(e) If the thread crumbles splitting transversely, and its surface is shiny from the beginning of rolling then the soil is very cohesive ($J > 30\%$); threads of less than 1 mm in diameter can be formed in very cohesive clays of sufficiently high water contents.

The 'soaking' test can be used as an additional check of the clay fraction content in soils. In this simple test a small lump of dry cohesive soil is immersed in water: if the soil is a slightly clayey silty sand or silt, then it will soak and disintegrate immediately on immersion in water but, as the clay fraction content increases, so does the time required for the soil to soak and disintegrate.

The content of sand and silt fractions in cohesive soils is determined from the 'rubbing' test in which a pat of soil is rubbed between fingers immersed in water. Soils containing over 50% of sand fraction leave a large number of sand grains between the fingers and are definitely gritty in texture. Soils containing over 50% of silt fraction have a rough texture, but not gritty, and will leave only a few or no sand grains between fingers; very cohesive soils have a distinctly greasy feel.

The presence of silt fraction and fine sand can also be identified by the 'dilatancy' test: a pat of moist soil is shaken horizontally in the palm of the hand. If the silt content is high (and clay content low), then the surface of the pat will appear distinctly moist, but the moisture can be made to recede by pressing the pat with a finger; the dilatancy effect rapidly disappears as the clay content increases.

The type of soil is described according to Table 3.8.

Silty soils have generally been sedimented in still waters, e.g. lacustrine deposits, or have been deposited by wind, e.g. loess.

Cohesive sandy soils are generally glacial in origin and contain rock fragments in the form of rounded cobbles or gravel-size grains.

Glacial soils generally contain calcium carbonate and, on treatment with 20% solution of hydrochloric acid, liberate carbon dioxide which is observable as extensive bubbling of the acid solution.

Estuarine soils (alluvial muds) are generally dark in colour and require a large number of thread-rolling operations (over 10) before crumbling. Peaty soils or peats generally contain some semi-decayed organic matter.

Determination of consistency of soils. The consistency index of cohesive soils is determined from the number of rolling cycles in the thread-rolling test; if the thread crumbles, for example, at the fourth rolling attempt, then it is noted that the number of rolling cycles is three (because already at the beginning of the fourth rolling of the thread the water content was close to the plastic limit).

The consistency index is determined from Equation (3.25):

$$I_c = 1 - \frac{1 \cdot 25N}{AJ}$$

Table 3.8. Identification of type of cohesive soils

	Group of soil			Recognition of cohesiveness of soil	
	Sandy	Intermediate	Silty		
Cohesiveness of soil (active clay fraction J (%))	sand fraction $> 50\%$ silt fraction $< 30\%$	sand fraction $> 30\%$ silt fraction $> 30\%$	sand fraction $< 30\%$ silt fraction $> 30\%$	Thread-rolling test on moist soil	Soaking test on dry soil
	Rubbing test in water				
	large number of sharp sand grains present	individual small sand grains present	sand grains cannot be felt but coarse texture		
very slightly cohesive $J < 5\%$	slightly clayey sand (positive dilatancy in fine sand)	sandy silt (positive dilatancy)	silt (positive dilatancy)	thread flattens or crumbles– cannot be rolled (Figure 3.20)	soaks and disintegrates immediately
slightly cohesive $J = 5–10\%$	slightly clayey sand (some dilatancy in fine sand)	sandy silt (some dilatancy)	silt (some dilatancy)	thread crumbles, splits longitudinally like a cigar (Figure 3.21)	soaks and disintegrates in few minutes
medium cohesive $J = 10–20\%$	clayey sand	clayey sandy silt	clayey silt	thread splits transversely; remains dull, matt, during rolling (Figure 3.22)	takes a long time to soak
cohesive $J = 20–30\%$	sand–clay	sand–silt–clay	silt–clay	thread splits distinctly transversely, becomes shiny towards end of rolling	
very cohesive $J > 30\%$	sandy clay	clay	silty clay	small ball and thread shiny from beginning of rolling, greasy feel	

Notes. Dry lumps of fine sand and silt can be powdered between fingers while dry lumps of clay can be broken but not powdered. Wet or moist silts leave a powdery deposit on the fingers which *dries* quickly and can then be brushed off; clayey soils ($J > 10\%$) and clays stick to the fingers, dry slowly, and then have to be washed off.

where 1·25 = amount of water in % that the thread loses during a single rolling cycle

N = number of rollings from natural water content until crumbling occurs at the plastic limit

A = colloidal activity: for most glacial clays approximately 1·0, for late glacial and post-glacial lacustrian clays and loesses A = 0·4 to 0·75, and for organic estuarine clays $A > 1{\cdot}25$

J = average clay fraction content in % (from Table 3.8)

Example. During the thread-rolling test on a boulder clay it was observed that the thread split transversely and that its surface became shiny towards the end of the test.

It is concluded from the above that the soil is 'cohesive' with an average clay content J = 25%. Since the rubbing test revealed a considerable amount of sand the soil is of the *sand–clay* type (Table 3.8). For a soil of glacial origin A = 1·0. The number of thread-rolling cycles was equal to 3, i.e. N = 3. Substituting into Equation (3.25),

$$I_c = 1 - \frac{1{\cdot}25 \times 3}{1 \times 25} = 1 - 0{\cdot}15 = 0{\cdot}85$$

Therefore the boulder clay can be more precisely described as *stiff* ($1{\cdot}0 \geqslant I_c > 0{\cdot}75$) *sand–clay*.

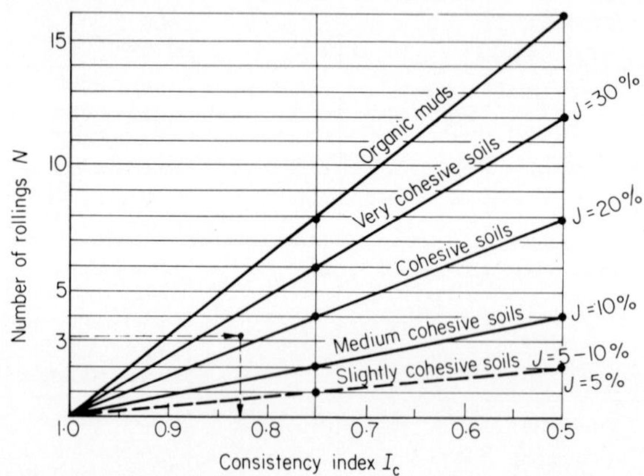

Figure 3.23. Nomogram for determination of consistency index for cohesive soils (having A = 1·0) from the thread-rolling test (see Table 3.5).

Consistency of soils for which activity can be approximately taken as equal to unity can also be determined from Figure 3.23.

3.12.3. STRUCTURAL CHARACTERISTICS OF COHESIVE SOILS

The existence of structural features in cohesive soils may influence their physical and mechanical properties and it is therefore essential that they should be included in the description of the soils (British Standard Code of Practice CP 2001, 1957).

Many stiff and harder clays exist in their natural state with a network of joints or fissures. When a large piece of such clay is dropped it breaks into polyhedral fragments; such clays are described as *fissured*. Clays deformed and folded by the weight of advancing and retreating glaciers, by tectonic activities, or by previous mass movements may contain discontinuities in the form of *slickensides.*

Within the zone of clay affected by near-surface weathering due to a seasonal wetting and drying fissuring is accentuated by the development of a columnar structure. At the same time softening of the clay takes place along the fissures due to the more rapid movement of water through them; such clays are described as *weathered.*

Both of the above structural characteristics are of prime importance in the choice of sampling techniques and in the method of determination of the mechanical properties of such clays.

Most of the soils of plastic consistency ($I_c < 0{\cdot}75$) have no apparent fissures and are described as *intact.* This is not affected by the presence of a few inclusions such as fossil shells or some organic matter.

Clays formed in glacial lakes usually have a finely laminated structure consisting of layers of clays (deposited in winter) and silts or fine sands (deposited in summer); clays are usually darker in colour. The silty layers act as partings and enable the material to be split up by hand into thin laminae. Such clays are described as *varved.*

Other cohesive soils which have laminations, more or less parallel with one another and with bedding planes, are referred to as *laminated* clays.

Laminated clayey or silty deposits which have been compacted naturally and have developed a fissility along the bedding planes are referred to as *shales.*

Soils consisting of alternating layers of varying types are described as *stratified* (if layers are thin the soils are described as laminated or varved—see above).

Soils consisting essentially of one type are described as *homogeneous.*

3.12.4. RECOGNITION OF TYPE AND DEGREE OF COMPACTION OF COHESIONLESS SOILS

The cohesionless soils can easily be distinguished from the cohesive soils because from the fine sand fraction upwards the individual grains are visible to a naked eye and when dry they do not form cohesive lumps; cohesionless soils cannot be rolled into threads. Simple directives for recognition and description of the type of cohesionless soils are given in Table 3.9.

The state of compaction of cohesionless soils can be established approximately in the field by driving in a 50 mm (2 in) square wooden peg: in loose deposits it can easily be driven in, while in dense deposits it is hard to drive it to a depth of more than 50 to 100 mm (2 to 4 in); in proper site investigation work either the dynamic or the static penetration tests are used for this purpose (Chapter 6).

Table 3.9. Identification of type of cohesionless soils

Type of soil	Content of grain (%)			Additional remarks
	>2 mm	>0·6 mm	>0·2 mm	
gravel	>50			
hoggin, till	50–10	>50		
coarse sand	<10	>50		individual grains are visible from a distance of a few metres
medium sand	<10	<50	>50	individual grains are visible from a distance of 1 m
fine sand	<10	<50	<50	individual grains are visible from a distance of 0·2 to 0·3 m
silty sand	<10	<50	<50	very fine sand; when dry forms slightly cohesive lumps, which when lifted disintegrate; powdery deposit remains on the fingers

3.12.5. RECOGNITION OF TYPE OF ORGANIC SOILS

Organic soils and peats are easily recognized by their colour and frequently by the odour; colours range from dark grey or brown to almost black and if fresh samples are inspected there is usually an unmistakable odour of decaying matter.

Soils containing less than about 50% of organic (vegetable) matter are referred to as 'organic' and are classified according to the characteristics of their inorganic fraction, i.e. within the groups of cohesionless or cohesive soils; the subsidiary organic constituent is included in the description in the following terms:

. . . with a trace . . .	organic content up to 10%,
. . . with a little . . .	organic content between 10 and 25%,
. . . with some . . .	organic content between 25 and 40%,
. . . and . . .	organic content about 50%.

Example. Soft dark-grey organic *sandy silt* with some fibrous peat, soft grey/black organic *silt* and amorphous organic matter.

Soil containing more than 50% of organic matter or pure organic soils are referred to as peats; the subsidiary constituents are included in the description in the manner described above but with appropriate changes in the wording.

The type of peat or organic content of a soil is described according to its macroscopic structure as shown in Table 3.10.

Table 3.10. Identification of type of peat and organic matter

Type of peat or organic matter	Appearance of freshly broken surface	Texture	Shrinkage on drying
fibrous	original plant structure visible	fibrous	limited but loosening of the texture
pseudo-fibrous	original plant structure visible; darkens on exposure	soft and plastic, can be squeezed through the fingers	appreciable
amorphous	fine grained—original plant structure not visible	soft, plastic, and greasy, can be squeezed through, fingers	appreciable—may break down into powder and angular fragments, when paddled can dry to form hard lumps
intermediate and mixed	mixture of more resistant fibrous matter in a strongly altered matrix	soft and plastic with fibrous matter or vice versa	generally appreciable

3.12.6. DESCRIPTION OF COLOUR

The colour of soil is described from inspection of the surface of a freshly broken lump of soil of natural water content. The basic colour is supplemented with a description of its shade and intensity; e.g. a grey soil of light yellow shading is described as light yellow/grey (the basic colour appearing at the end of the description).

3.12.7. STATES OF SATURATION

In the field, the state of saturation of soils is described as follows.

(a) *Dry.* If the soil does not moisten the fingers or in the case of a cohesive soil it does not deform plastically under pressure.

(b) *Damp.* If the soil does not moisten the fingers but in the case of a cohesive soil it deforms plastically.

(c) *Moist.* Paper or fingers which come in contact with the soil become damp.

(d) *Wet.* On squeezing in the hand the soil exudes water.

3.12.8. DETERMINATION OF CALCIUM CARBONATE CONTENT

The calcium carbonate ($CaCO_3$) content of soil can be determined approximately from the intensity of the reaction when a drop of a 20% solution of hydrochloric acid (HCl) is poured on it; in the case of a very intensive reaction (bubbling of the solution due to release of gas) several drops should be sprinkled on the soil.

If the solution

(a) bubbles intensively for a long time, then $CaCO_3$ content is over 5%,

(b) bubbles intensively for a short time, then $CaCO_3$ content is between 3 and 5%,

(c) bubbles slightly for a short time, then $CaCO_3$ content is between 1 and 3%, and

(d) bubbles very slightly, or does not bubble, then $CaCO_3$ content is less than 1%.

4

Flow of Water Through Soils

4.1. Types of Water in Soils

Water in the soil is found in the form of (a) adsorbed or hygroscopic water, strongly held on the surface of soil particles, (b) capillary water, supported in the pores by the surface tension forces above the free ground water table, and (c) free or gravitational water.

The adsorbed water is under the action of large molecular attraction forces and therefore is not free to flow under the gravitational forces.

The capillary water flows downwards when its weight exceeds the capillary surface tension forces.

The free water is completely subject to the gravitational forces and therefore occupies the lowest possible position in the pores of the permeable soils. The free or the gravitational water is commonly referred to as the ground water.

4.2. Ground Water

The ground water is found in underground troughs and beds filled with gravels and sands, i.e. in the soils more permeable than the underlying massive rock or clay formations.

The ground water is replenished through percolation of the rain water, through seepage of water from open reservoirs and rivers, and through condensation of water vapour present in the pores (Figure 4.1).

In its turn, the ground water appears on the ground surface in the form of springs and through seepage supplies some of the open water reservoirs.

The flow (or seepage) of the ground water always takes place towards the lower piezometric level (Figure 4.2). *The piezometric level* is defined as the final stable level to which water rises in a piezometric tube installed in a given stratum; a water-tight casing in a borehole can be considered as a piezometric tube installed in the water bearing stratum that it has reached.

There are two main types of the ground water: (a) normal, and (b) perched.

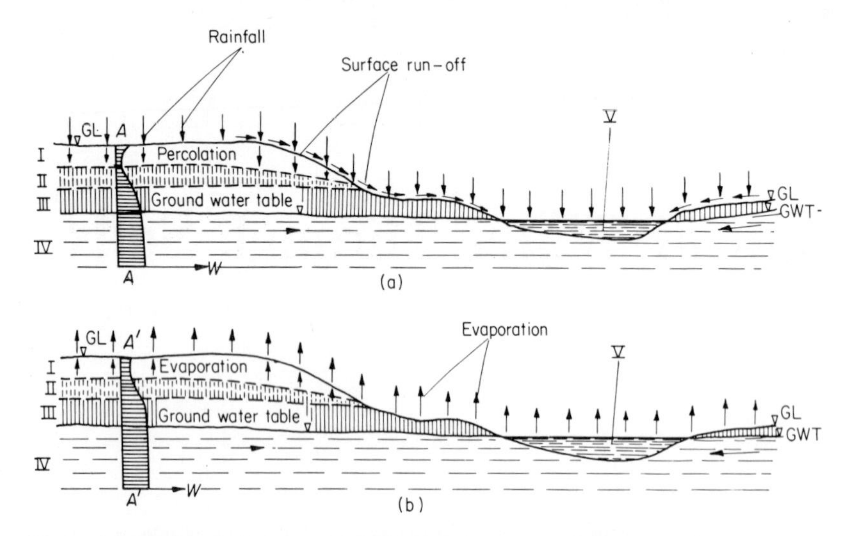

Figure 4.1. Ground water and open water levels: (a) increase during rainfalls; (b) lowering during drought; I, zone of percolation of rain water or evaporation of ground water; II, partly saturated capillary zone; III, fully saturated capillary zone; IV, zone of ground water; V, open water; A–A, distribution of natural water content during rainfall; A′–A′, during drought.

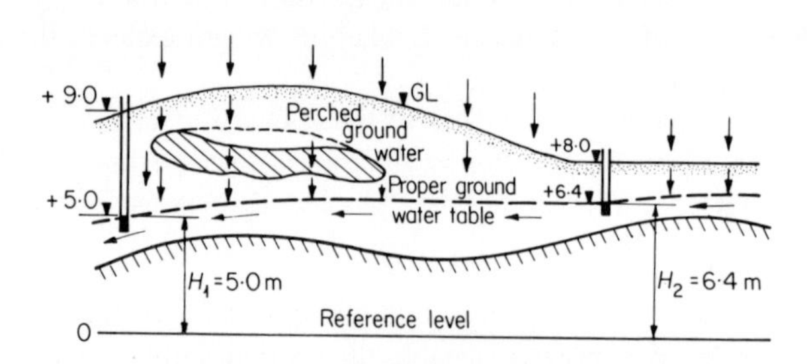

Figure 4.2. Perched and normal ground water.

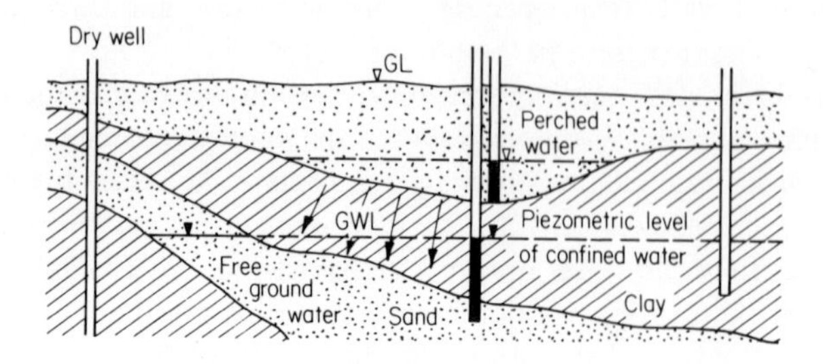

Figure 4.3. Perched, normal (or free), and confined ground water.

The normal ground water has a continuous surface profile (water table or phreatic surface) in water-bearing soils which usually extends over a large area; along any vertical line within the normal ground water the piezometric pressures increase linearly. On the other hand a *perched ground water* is found in local lenses contained by soils of low permeability and situated close to the ground surface, above the normal ground water level (Figures 4.2 and 4.3). The level of the perched ground water is greatly influenced by the amount of rainfalls (in the period before any investigation) and by the temporary periods of snow thawing.

Ground water can also be found confined between two strata of low permeability and in such a case will not have a free surface but will exert a pressure on the underside of the upper confining stratum (Figure 4.3). This type of ground water is referred to as the sub-Artesian or confined ground water and the permeable stratum in which it is found is known as a confined aquifer.

Conditions may exist in which the upper confining stratum follows the ground profile and then thins out as shown in Figure 4.4; the piezometric level

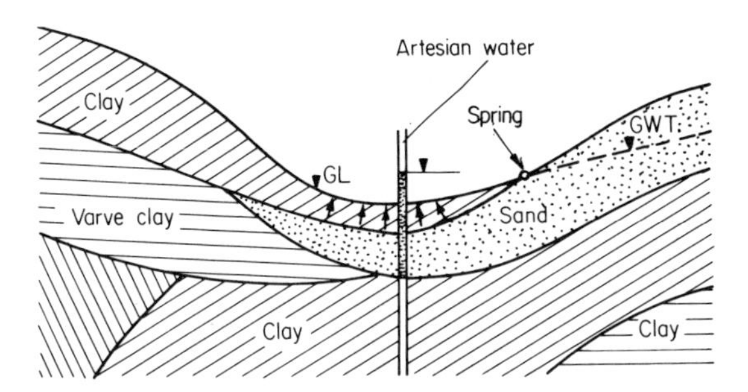

Figure 4.4. Artesian water.

of the confined ground water may then be above the ground level; such water is referred to as the Artesian water.

The flow of the confined and Artesian water also takes place towards the lower piezometric level.

An accurate knowledge of the ground water conditions (types of ground water and their levels) is of considerable practical importance, because it enables the determination of the directions and the rates of flow of the different types of water and prediction of their influence on the behaviour of the ground beneath a structural foundation.

The knowledge of the ground water level can also be used to check if a given interpretation of the geological structure of the area is correct, e.g. a confined continuous permeable stratum (confined *aquifer*) should have continuous and reasonably smooth piezometric levels.

4.3. Permeability of Soils

The capacity of soil to allow water to pass through it, through a network of channels formed by its pores, is known as the *permeability*. The resistance that the soil offers to the flow of water mainly depends on the following.

(a) Particle-size distribution.
(b) Porosity.
(c) Mineralogical composition of solids and the type of the adsorbed cations.
(d) Temperature of water (viscosity).

In normal circumstances the flow of water in soils is induced by gravitational forces, which tend to equalize the difference in water levels in two reservoirs between which the flow of water can take place. In practice, any soil within which the flow of water takes place can be considered as a system of interconnected reservoirs. The force inducing the flow is the product of the difference in the water levels and its unit weight. The force is distributed uniformly along the whole length of the flow path (if the resistance to flow

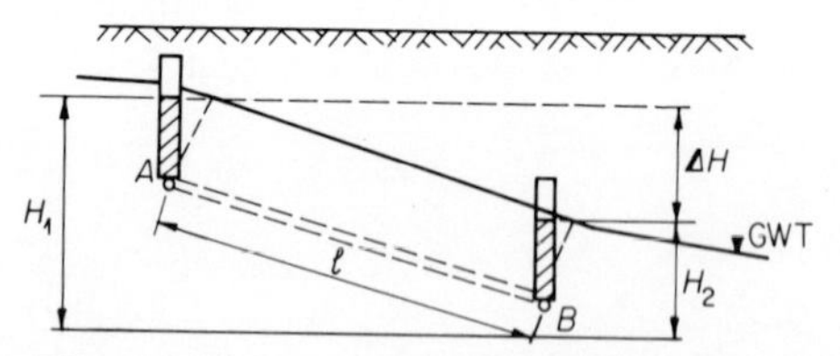

Figure 4.5. Determination of hydraulic gradient.

along the whole flow path is uniform), and hence the rate of flow of water will depend on the ratio of the excess head to the length of the flow path, i.e. on the hydraulic gradient (Figure 4.5):

$$i = \frac{\Delta H}{l} \tag{4.1}$$

where i = hydraulic gradient (dimensionless)
ΔH = difference in piezometric water levels in m
l = length of low path in m

The measure of permeability of soils is the coefficient (of permeability) 'k', also known as Darcy's constant, which relates the rate of flow of water through the soil v to the hydraulic gradient i:

$$v = ki \tag{4.2}$$

where v = rate of flow of water in mm/s (ft/s, cm/s)
k = coefficient of permeability—Darcy's constant, having the dimensions of velocity, e.g. mm/s (cm/s); coefficient of permeability k is equal to the rate of flow v for $i = 1$
i = hydraulic gradient (dimensionless)

The value of the coefficient k is characteristic of a given soil medium, i.e. it does not depend on the hydraulic gradient but depends on the porosity of the soil, on its particle-size distribution, and on the temperature of the flowing water. The coefficient k for fine-grained soils (sand–clays, silt–clays, clays) can be as low as 10^{-9} mm/s while for gravels it can exceed 1 mm/s. Typical values of the coefficient of permeability of different soils are given in Table 4.1.

Table 4.1. Average values of coefficient of permeability

Type of soil	Coefficient of permeability k (mm/s)
fine gravel	100–1
medium and coarse sand*	$1-10^{-1}$
fine sand	$10^{-1}-10^{-2}$
silty sand	$10^{-2}-10^{-3}$
loess of undisturbed structure	$10^{-2}-10^{-3}$
loess remoulded	$10^{-4}-10^{-6}$
silts and slightly clayey sands	$10^{-3}-10^{-5}$
clayey soils	$10^{-5}-10^{-7}$
sand–clays, silt–clays	$10^{-6}-10^{-8}$
clays	$10^{-7}-10^{-9}$

* For river sands and sandy gravels k can be taken as 0·25 and 0·75 mm/s respectively.

In *approximate* evaluation of the coefficient of permeability k of cohesionless soils Hazen's formula is frequently used:

$$k = Cd_{10}^2 \tag{4.3}$$

where k = Darcy's constant in mm/s
C = experimental coefficient: for clean uniformly graded sands it varies between 12 and 8; for slightly clayey silty sands and for non-uniformly graded sands between 8 and 6
d = effective size or diameter (Section 3.5) in mm

It is generally taken that Hazen's formula is applicable to sands whose effective diameters d_{10} vary between 0·1 and 3·0 mm and whose coefficient of uniformity U is not greater than 5. Experiments carried out by Laudon (1952) and Golder and Gass (1962) suggest, however, that this range can be extended down to about d_{10} = 0·003 mm and to non-uniform and well-graded soils ($U \gg 5$); for such soils the experimental coefficient C should be taken as approximately equal to 5·0 while the effective diameter d_{10} should be determined statistically from a number of samples of the same soil.

An empirical relationship between the coefficient k and the voids ratio e for cohesionless soils is of the following form:

$$k = C_1 e^2 \tag{4.4}$$

where k = Darcy's constant
C_1 = experimental coefficient
e = voids ratio

The effect of temperature on the rate of flow of water through soils can be explained by the fact that the viscosity of water is very sensitive to temperature change; an empirical relationship between the coefficient k and water temperature is given by Equation (4.5). For practical purposes the value of k determined experimentally at a given temperature T is reduced to the value corresponding to the average temperature of the ground water of +10 °C, i.e. to k_{10}:

$$k_{10} = \frac{k_T}{0{\cdot}7 + 0{\cdot}03T} \tag{4.5}$$

where T = temperature of the water in °C.

So far, only the simple linear velocity of flow has been considered; in practical problems, as well as in most of the laboratory tests, the knowledge of the actual quantity of flow (or volumetric rate of flow) is of more significance and can be directly measured in the laboratory. A simple relationship exists between the quantity of flow, the coefficient k, and the hydraulic gradient:

$$Q = k_T i A t \tag{4.6}$$

where Q = quantity of flow
k_T = Darcy's constant (for water temperature T °C)
A = cross-sectional area of soil over which flow takes place, normal to the direction of flow
t = time
i = hydraulic gradient (dimensionless)

In the above equation the values of Q, t, A, and the hydraulic gradient i are usually obtained from laboratory tests and hence, for a given temperature T, the coefficient k can be evaluated as follows:

$$k_T = \frac{Q}{Ati} \tag{4.7}$$

4.4. Basic Directions of Flow of Water in the Ground

The following model arrangements illustrate the three basic directions of flow of water in the ground.

(a) In a horizontal direction (Figure 4.6).
(b) In a vertical direction, downwards (Figure 4.7).
(c) In a vertical direction, upwards (Figure 4.8).

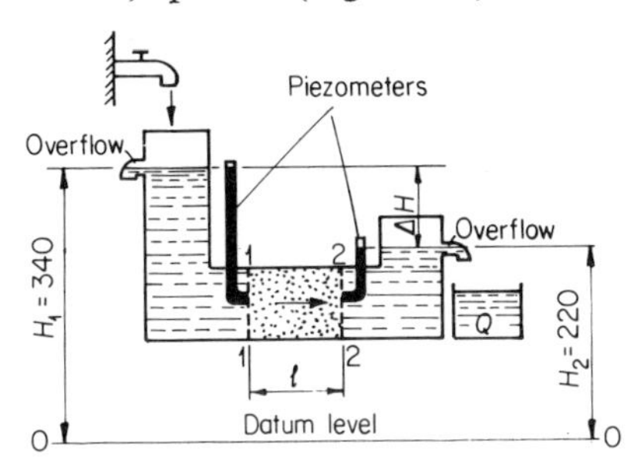

Figure 4.6. Horizontal flow of water.

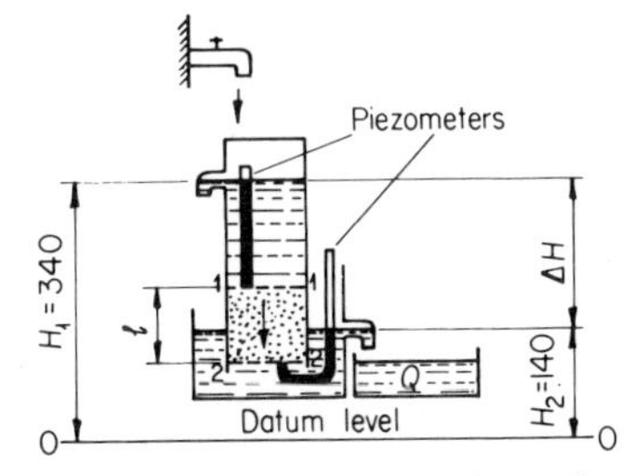

Figure 4.7. Vertical downward flow of water.

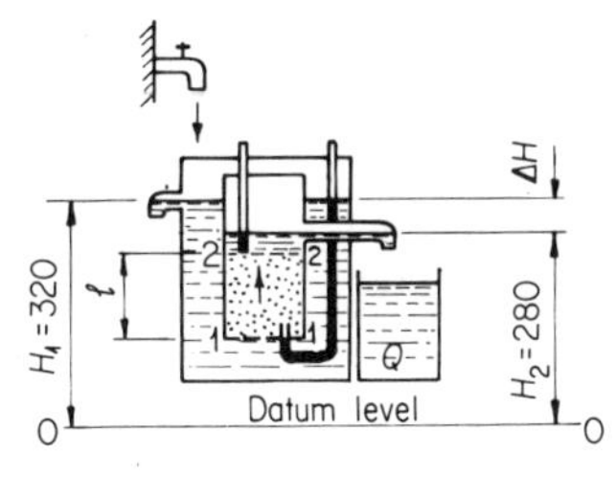

Figure 4.8. Vertical upward flow of water.

The analysis of flow of water in laboratory models, as well as in the ground, can be greatly simplified by adopting the following procedure.

(1) Establish the direction of flow (indicated by arrows).

(2) Establish section 1–1 at which the water enters the soil layer and section 2–2 at which it leaves it.

(3) Determine the piezometric heads (levels) H_1 and H_2 for sections 1–1 and 2–2, respectively, with reference to datum level 0–0.

(4) Determine the difference in piezometric levels $\Delta H = H_1 - H_2$.

(5) Determine the length of the flow path l.

(6) Determine the hydraulic gradient $i = \Delta H/l$.

(7) Determine the cross-sectional area A over which the flow takes place.

The above data enable the quantity of flow Q to be determined from Equation (4.6), if the coefficient of permeability k is known; conversely, the coefficient k can be determined if Q is known.

Flow of water in a horizontal direction. The model illustrated in Figure 4.6 corresponds to the frequently encountered horizontal flow of confined water in permeable formations sandwiched between impervious strata (Figure 4.2).

Example. In a laboratory test on a model illustrated in Figure 4.6 the quantity of flow recorded over 30 minutes was $Q = 640$ ml. Determine the coefficient of permeability k.

In accordance with the above-described procedure (steps (1)–(7)) the data required for evaluation of k are indicated in Figure 4.6.

Substituting these data into Equation (4.7) the coefficient of permeability k is obtained and using Equation (4.5) k_{10} can be evaluated:

$$H_1 = 340 \text{ mm}, \qquad H_2 = 220 \text{ mm}, \qquad \Delta H = 340 - 220 = 120 \text{ mm}$$

$$l = 150 \text{ mm}, \qquad i = \frac{\Delta H}{l} = \frac{120}{150} = 0{\cdot}8$$

$$A = 8000 \text{ mm}^2, \qquad \text{water temperature } T = 22\ ^\circ\text{C}$$

time over which Q was recorded $t = 30$ min

$$k_T = \frac{Q}{iAt} = \frac{640 \times 10^3}{0{\cdot}8 \times 8000 \times 30} = 3{\cdot}3 \text{ mm/min}$$

$$k_{10} = \frac{k_T}{0{\cdot}7 + 0{\cdot}03T} = \frac{3{\cdot}3}{0{\cdot}7 + 0{\cdot}03 \times 22} = 2{\cdot}4 \text{ mm/min}$$

If k_{10} was known then a similar procedure would be adopted to determine the quantity of flow Q.

Flow of water vertically downwards. As the next model let us consider the vertically downward flow of water. This type of flow is most frequently encountered in seepage of water from a perched water table through the underlying stratum of low permeability (Figures 4.2 and 4.3).

Example. A hydraulic model shown in Figure 4.7 has been constructed using medium sand. Determine k_{10} for $t = 20$ min and $Q = 480$ ml.

Following the given procedure the necessary data are evaluated:

$H_1 = 340$ mm, $\quad H_2 = 140$ mm, $\quad \Delta H = 340 - 140 = 200$ mm

$l = 100$ mm, $\quad i = \dfrac{\Delta H}{l} = \dfrac{200}{100} = 2$

$A = 8000$ mm^2

For $t = 20$ min and $Q = 480$ ml, at $T = 15$ °C we obtain

$$k_T = \frac{480 \times 10^3}{2 \times 8000 \times 20} = 1{\cdot}5 \text{ mm/min}$$

$$k_{10} = \frac{1{\cdot}5}{0{\cdot}7 + 0{\cdot}03 \times 15} = 1{\cdot}3 \text{ mm/min}$$

The previous model can be utilized in evaluation of the quantity of water flowing vertically downwards through a stratum of clayey silt–sand (Figure 4.9).

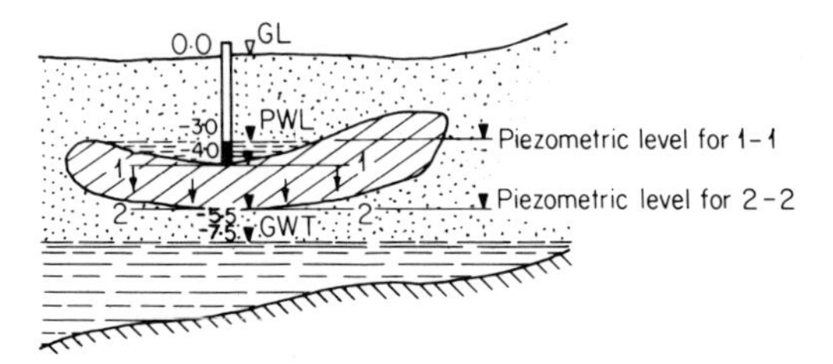

Figure 4.9. Downward seepage of water through the stratum of clayey silt–sand.

Example. Data are given in Figure 4.9. Determine the quantity of flow per one hour, given that $k_{10} = 10^{-7}$ mm/min.

Following the set procedure one establishes the direction of flow, the sections at which water enters and leaves the soil layer and the corresponding piezometric levels, the length of the flow path, and the hydraulic gradient.

Level of section at which water enters $1-1 = -4{\cdot}0$ m
Piezometric level for section $1-1 = -3{\cdot}0$ m
Level of section at which water leaves $2-2 = -5{\cdot}5$ m
Piezometric level for section $2-2 = -5{\cdot}5$ m
Length of flow path $l = -4{\cdot}0 - (-5{\cdot}5) = 1{\cdot}5$ m
Difference in piezometric levels $\Delta H = -3{\cdot}0 - (-5{\cdot}5) = 2{\cdot}5$ m
Hydraulic gradient $i = \dfrac{\Delta H}{L} = \dfrac{2{\cdot}5}{1{\cdot}5} = 1{\cdot}67$

The quantity of water seeping vertically downwards, in one hour, through an area of 1 m^2 is obtained from Equation (4.6):

$$Q = k_T i A t = 10^{-7} \times 1{\cdot}67 \times 10^3 \times 60 = 0{\cdot}01 \text{ ml}$$

Flow of water vertically upwards. The upward flow is frequently encountered in practice when the water is pumped from an excavation in a soil formation of low permeability, which is subject to the pressure of the confined water below it (Figure 4.10).

A laboratory model of such flow is shown in Figure 4.8.

Example. Data as given in Figure 4.8. Determine k_{10}.

Following the set procedure the necessary data are evaluated:

$$H_1 = 320 \text{ mm}, \quad H_2 = 280 \text{ mm}, \quad \Delta H = 320 - 280 = 40 \text{ mm}$$

$$l = 100 \text{ mm}, \quad i = \frac{\Delta H}{l} = \frac{40}{100} = 0{\cdot}4$$

$$A = 8000 \text{ mm}^2, \quad t = 10 \text{ min}$$

$$Q = 200 \text{ ml at } T = 15^\circ\text{C}$$

Therefore

$$k_T = \frac{200 \times 10^3}{0{\cdot}4 \times 8000 \times 10} = 6{\cdot}2 \text{ mm/min}$$

$$k_{10} = \frac{6{\cdot}2}{0{\cdot}7 + 0{\cdot}03 \times 15} = 5{\cdot}4 \text{ mm/min}$$

The above model and example can be utilized for determination of the quantity of water seeping through the bottom of the excavation as shown in the following example.

Example. Data as given in Figure 4.10. Determine the quantity of flow Q per one hour.

Following the set procedure, the direction of flow, the sections at which water enters and leaves the soil layer and the corresponding piezometric levels, the length of the flow path, and the hydraulic gradient are established.

Level of section at which water enters $1-1 = -5{\cdot}0$ m
Piezometric level for section $1-1 = -1{\cdot}2$ m
Level of section at which water leaves $2-2 = -3{\cdot}0$ m
Piezometric level for section $2-2 = -3{\cdot}0$ m
Length of flow path $l = -3{\cdot}0 - (-5{\cdot}0) = 2{\cdot}0$ m
Difference in piezometric levels $\Delta H = -1{\cdot}2 - (-3{\cdot}0) = 1{\cdot}8$ m
Hydraulic gradient $i = \Delta H/l = 1{\cdot}8/2{\cdot}0 = 0{\cdot}9$

The quantity of water seeping through 1 m^2 of the bottom of the excavation, in one hour, in the silt having $k_{10} = 2 \times 10^{-4}$ mm/s and at ground water temperature of 18 °C is

$$k_{18} = 2 \times 10^{-4}\,(0{\cdot}7 + 0{\cdot}03 \times 18) = 2 \times 10^{-4} \times 1{\cdot}24 = 2{\cdot}5 \times 10^{-4} \text{ mm/s}$$

$$Q = k_T iAt = 2{\cdot}5 \times 10^{-4} \times 0{\cdot}9 \times 10^3 \times 60 \times 60 = 810 \text{ ml}$$

The inflow of water through 1 m^2 of the bottom of the excavation, per one hour, is approximated equal to 1·0 litre.

4.5. Flow Net

In the case of flow of ground water under variable hydraulic gradient and depth of seepage zone the evaluation of the quantity of flow is not as simple as described in the above section and can only be determined with the help of a flow net.

Let us consider a two-dimensional 'steady-state' condition* of flow of water through a soil element of dimensions dx, dy, and dz (Figure 4.11). We shall

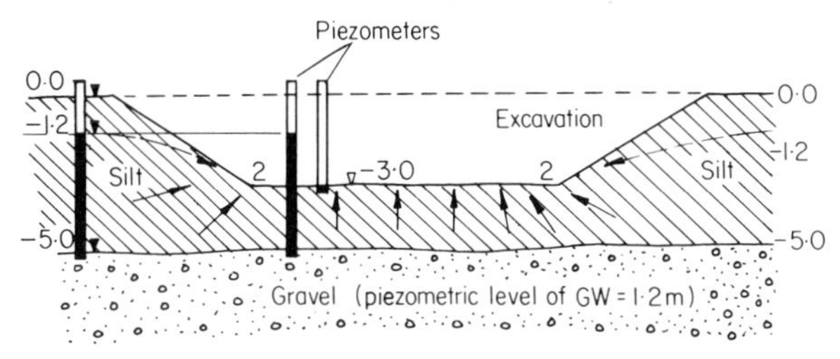

Figure 4.10. Upward seepage of water into excavation in silt.

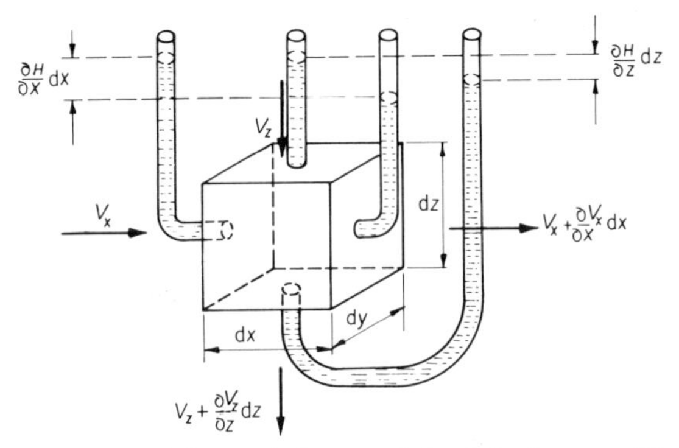

Figure 4.11. Velocities of flow and piezometric levels in the case of flow of water through a cubical element.

consider the case of the two-dimensional flow parallel to the (x, z) plane, i.e. normal to the y axis.

The velocity of flow in the horizontal direction is denoted by V_x and in the vertical direction by V_z.

The volumetric rate of flow of water into the element under consideration is

$$V_x \, dz \, dy + V_z \, dx \, dy$$

* Two-dimensional 'steady-state' condition of flow is defined as that in which there is now inflow or outflow of water across the sides parallel to the plane of flow of the element under consideration and the rate of flow is constant.

while the rate of flow of water out of the element is

$$V_x \, dz \, dy + \frac{\partial V_x}{\partial x} dx \, dz \, dy + V_z \, dx \, dy + \frac{\partial V_z}{\partial z} dz \, dx \, dy$$

Considering now that (1) the water and the soil skeleton (solid phase) are practically incompressible, i.e. their volume remains constant, (2) the flow is taking place under steady conditions, and (3) the cavitation does not occur in the water, we conclude that the quantity of water entering the element must be equal, in a given time, to the quantity flowing out of it.

Therefore

$$\frac{\partial V_x}{\partial x} dx \, dz \, dy + \frac{\partial V_z}{\partial z} dz \, dx \, dy = 0$$

or

$$\frac{\partial V_x}{\partial x} + \frac{\partial V_z}{\partial z} = 0 \tag{4.8}$$

In hydrodynamics the above equation is known as the differential equation of the continuity of flow.

According to Darcy's law both the velocity components can be expressed in terms of the coefficient k and the corresponding hydraulic gradients:

$$V_x = k \frac{\partial H}{\partial x} \quad \text{and} \quad V_z = k \frac{\partial H}{\partial z}$$

where $\partial H / \partial x$ and $\partial H / \partial z$ are the hydraulic gradients in x and z directions respectively.

The product of the coefficient of permeability and the difference in the piezometric levels between any two sections under consideration is known as the flow potential $\phi = k \Delta H$.

On differentiation and substitution we obtain

$$V_x = \frac{\partial \phi}{\partial x} \quad \text{and} \quad V_z = \frac{\partial \phi}{\partial z}$$

Substituting the above values in Equation (4.8) we obtain

$$\frac{\partial^2 \phi}{\partial x^2} + \frac{\partial^2 \phi}{\partial z^2} = 0 \tag{4.9}$$

Equation (4.9) is known as Laplace's equation and can be illustrated graphically by two families of curves intersecting each other at right angle. Such a graph is known as the *flow net.*

A flow net (Figure 4.12) is an orthogonal net (all lines intersect at right angle), constructed of flow lines (lines parallel to the direction of flow of water), and equipotential lines (lines intersecting the flow lines at right angles).

The equipotential lines are loci of points of the same piezometric level: the water will rise to the same level in piezometers placed at any point along an equipotential line (section 14 in Figure 4.12).

The construction of a flow net involves plotting of an orthogonal net within the longitudinal section of the seepage zone. The lines bounding the seepage

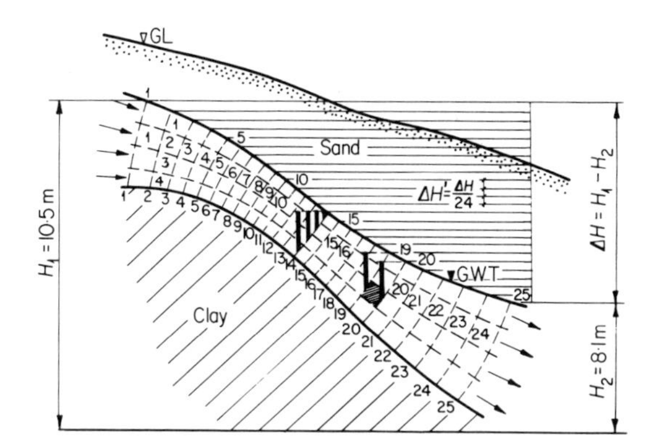

Figure 4.12. Flow net illustrating a downhill seepage of water.

zone in Figure 4.12 are from the top the ground water table and from the bottom the profile of the underlying impermeable stratum.

The equipotential lines are drawn to intersect the flow lines at right angles, and to form approximately square net elements: as these lines are drawn an attempt is made to keep the length of the elements equal to their height.

At the same time it must be ensured that the differences in piezometric levels between any two adjacent equipotential lines are equal ($\Delta H'$ = constant).

In order to determine the total quantity of flow we will consider the flow of water through the shaded element in Figure 4.12; average length and height of the element is taken as a. The hydraulic gradient across the element is equal to

$$i = \frac{\Delta H'}{a} = \frac{\Delta H}{am} \qquad \text{(a)}$$

where m = number of net elements in the direction of flow (in Figure 4.12, $m = 24$)

The rate of flow

$$v = ki = k\frac{\Delta H}{am} \qquad \text{(b)}$$

The quantity of water flowing through one element of width $b = 1$ is

$$\Delta Q = vab = k\frac{\Delta H}{am}a = k\frac{\Delta H}{m} \qquad \text{(c)}$$

The total quantity flowing through the entire depth of the seepage zone is

$$Q = n \times \Delta Q = k\Delta H \frac{n}{m} \tag{4.10}$$

where n = the number of elements in the direction at right angle to the direction of flow (in Figure 4.12, $n = 4$).

Equation (4.10) enables the determination of the quantity of flow of water on the basis of a simply constructed flow net.

A flow net also enables us to determine the hydraulic gradient at any point within the seepage zone.

Example. Data as in Figure 4.12:

$$H_1 = 10{\cdot}5 \text{ m}, \quad H_2 = 8{\cdot}1 \text{ m}, \quad k = 10^{-1} \text{ mm/s}$$

Determine the quantity of water flowing per minute through a cross section of width $b = 3{\cdot}0$ m.

$$Q = k\Delta H \frac{n}{m} b = \frac{10^{-1} \times 60}{1000}(10{\cdot}5 - 8{\cdot}1)\frac{4}{24} \times 3 = 7{\cdot}2 \times 10^{-3} \text{ m}^3/\text{min}$$

$$= 7{\cdot}2 \text{ l/min}$$

Electrical analogue. Because of the work and difficulties involved in plotting a flow net by the 'trial and error' method, particularly in the case of complicated boundary profiles, the method of electrodynamic analogy is frequently used in practice. The method is based on the analogy between the differential

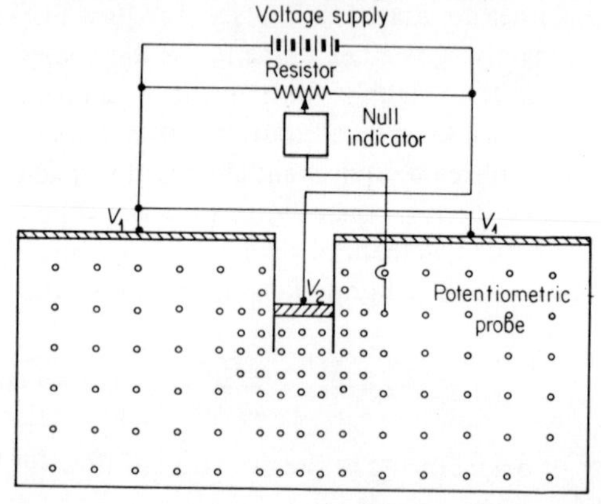

Figure 4.13. Typical layout of an electrical analogue.

equations which govern the flow of electric current and flow of water in homogeneous isotropic media.

A typical layout of an electrical analogue is shown in Figure 4.13. In electrical models voltage corresponds to the piezometric head, conductivity to permeability, and current to velocity. By measuring voltage at a number of points the equipotential lines can be sketched (with some types of apparatus these can be plotted directly by finding loci of points of constant voltage drop). To complete a flow net, flow lines are sketched in at right angles to the equipotential lines.

4.6. Seepage Forces

Flow of water through soils induces frictional resistance between the water and the soil particles. From consideration of equilibrium of a small element of soil it is obvious that this resistance must be equal to the difference of the hydrostatic water pressures acting on the two faces of the element normal to the flow (e.g. sections 19 and 20 in Figure 4.12).

It can be seen from Figure 4.12 that the difference in the hydrostatic water pressures acting on a cubical element of side a is

$$p = a^2 \Delta H' \gamma_w = a^2 \frac{a\Delta H'}{a} \gamma_w = a^3 i \gamma_w \qquad \text{(d)}$$

Therefore, the force exerted by the water pressure on soil particles contained in a unit volume of the soil is

$$p_s = \frac{P}{V} = i\gamma_w \qquad (4.11)$$

This force is known as the seepage (hydrodynamic) force; it is equal in magnitude to the product of the hydraulic gradient 'i' and the unit weight of water γ_w. The direction of action of this force is tangential to the flow line; it is a force per unit volume and hence has the same units as γ_w—if g (gravitational acceleration) is in m/s^2 and ρ_w (density of water) in t/m^3 then its units are kN/m^3.

4.7. Effect of Seepage Force on the Unit Weight of Soil

In the case of vertical seepage of water through the soil it is necessary to include the effect of seepage force on the unit weight of soil γ; considering that the soil is below the ground water table its submerge unit weight is

$$\gamma' = \gamma_{sub} \pm p_s = g(\rho_{sat} - \rho_w) \pm p_s \qquad (4.12)$$

where ρ_{sat} = saturated density of soil in t/m^3
ρ_w = density of water in t/m^3
g = gravitational acceleration in m/s^2
γ_{sub} = submerged unit weight of soil in kN/m^3
p_s = vertical seepage force* in kN/m^3

The positive sign refers to the downward flow of water (Figure 4.14, element 3); the negative sign refers to the upward flow of water (Figure 4.14, element 20).

* In the case when the direction of seepage is inclined at an angle α to the vertical then only the vertical component of vector p_s is considered: $\overline{p_s} = p_s \cos \alpha$.

Example. Data: excavation of dimensions shown in Figure 4.14, homogeneous isotropic soil, sand of saturated density $\rho_{sat} = 1{\cdot}91$ t/m^3. Determine seepage unit force and the effective unit weight of sand in elements 3/1 and 20/4.

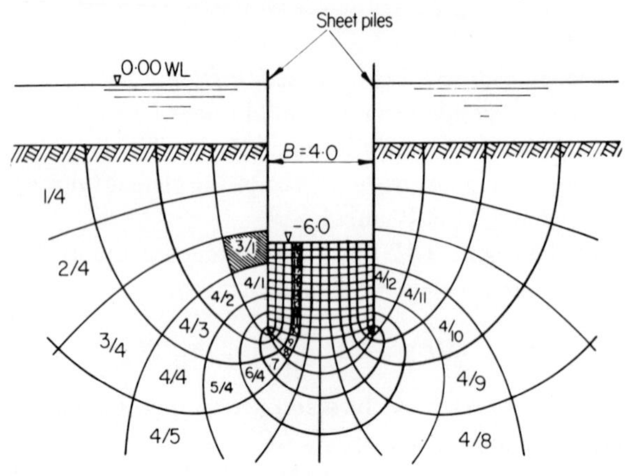

Figure 4.14. Flow net for seepage of water into excavation between two lines of sheet piles.

Seepage unit force $p_s = \Delta H g \rho_w / ma$, effective unit weight $\gamma' = \gamma_{sub} + p_s$. Determination of p_s and γ' for element 3/1:

$$\Delta H = 6{\cdot}0 \text{ m}, \quad m = 20, \quad a_3 = 1{\cdot}6 \text{ m}$$

$$\gamma_{sub} = g(\rho_{sat} - \rho_w) = 9{\cdot}81\ (1{\cdot}91 - 1{\cdot}0) = 8{\cdot}93 \text{ kN/m}^3, \ \Delta H' = \frac{\Delta H}{m}$$

$$l_3 \approx a_3, \ p_s = i\gamma_w = \frac{\Delta H}{m}\frac{\gamma_w}{a_3}$$

Hence

$$p_s = \frac{6{\cdot}0 \times 9{\cdot}81}{20 \times 1{\cdot}6} = 1{\cdot}86 \text{ kN/m}^3$$

and

$$\gamma' = 8{\cdot}93 + 1{\cdot}86 = 10{\cdot}89 \text{ kN/m}^3$$

Determination of p_s and γ' for element 20/4:

$$\Delta H = 6{\cdot}0 \text{ m}, \ m = 20, \ l_{20} \approx a_{20} = \frac{B}{n} = \frac{4{\cdot}0}{12} \approx 0{\cdot}33 \text{ m}$$

Hence

$$p_s = \frac{6{\cdot}0 \times 9{\cdot}81}{20 \times 0{\cdot}33} = 8{\cdot}93 \text{ kN/m}^3$$

and

$$\gamma' = 8{\cdot}93 - 8{\cdot}93 = 0$$

As can be seen, the effective unit weight of sand in the bottom of the excavation between the sheet piles is reduced by the upward seepage force to zero which will undoubtedly lead to its loosening and to the 'quick-sand' condition.

In the case of layered soils of considerably differing permeabilities it can be assumed that the seepage force is wholly transmitted to the less permeable soil.

This can be demonstrated by a simple analysis of the conditions of flow in the apparatus shown in Figure 4.15.

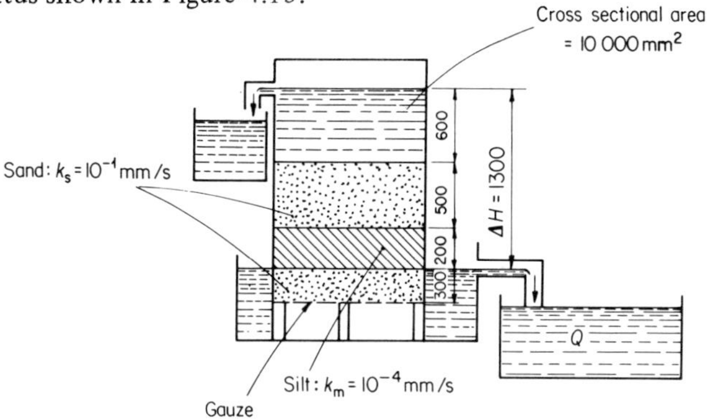

Figure 4.15. Flow of water through layered soils.

The quantities of water flowing through the sand and silt must be equal in any interval of time. Therefore, the rates of flow of water through the sand V_s and silt V_m must also be equal, because the cross-sectional area is constant throughout.

That is, because $V = ki$, we can write that

$$k_s i_s = k_m i_m$$

The hydraulic gradient in the silt is equal to

$$i_m = \frac{k_s}{k_m} i_s$$

Because the ratio of the coefficients $k_s : k_m$ for these materials is equal to 1000, therefore the hydraulic gradient in the silt is 1000 times greater than the hydraulic gradient in the sand; it can be assumed that the seepage force is wholly taken up by the silt.

Example. Data as in Figure 4.10.

Coefficient of permeability of silt	$k = 10^{-4}$ mm/s
Coefficient of permeability of gravel	$k = 1$ mm/s
Saturated density of silt	$\rho_{sat} = 1{\cdot}9$ t/m^3
Submerged unit weight of silt	$\gamma_{sub} = g(\rho_{sat} - \rho_w)$
	$= 9{\cdot}81(1{\cdot}9 - 1{\cdot}0) = 8{\cdot}83$ kN/m^3

Determine the resultant force on the silt in the bottom of the excavation due to an upward seepage of water (from the gravel). Because of the considerably differing permeabilities of the gravel and silt we can assume that the whole of the seepage force is transmitted to the silt layer. According to Equation (4.12)

$$\gamma' = \gamma_{sub} - p_s$$

where

$$p_s = i\gamma_w = ig\rho_w$$

From data in Figure 4.10

$$i = \frac{-1{\cdot}2 - (-3{\cdot}0)}{-3{\cdot}0 - (-5{\cdot}0)} = 0{\cdot}9$$

and therefore

$$p_s = 0{\cdot}9 \times 9{\cdot}81 = 8{\cdot}83 \text{ kN/m}^3$$

and

$$\gamma' = 8{\cdot}83 - 8{\cdot}83 = 0$$

The unit weight of the silt has been reduced to zero; in such conditions loosening and heave of the soil in the bottom of the excavation will take place owing to the upward seepage of water.

The seepage force equal to the unit submerged weight of the soil is known as the *critical seepage force*; for the above example, $p_{crit} = 8{\cdot}83$ kN/m^2, at a hydraulic gradient $i_{crit} = 0{\cdot}9$.

As can be seen the upward seepage force can have a destructive effect on the soil. When excavating in sands the following condition should be satisfied:

$$p_s \leqslant \frac{\gamma_{sub}}{n} \tag{4.13}$$

where the factor of safety $n \geqslant 1{\cdot}5$.

In order to prevent local heave and piping in layers of low permeability (silts, silty sands, and slightly clayey sands) the factor n should be taken as equal to or greater than 2·5. In the presence of layers of soils susceptible to swelling it is recommended that the piezometric level of water should be lowered below the bottom of the excavation.

4.8. Laboratory Methods of Determination of Coefficients of Permeability of Soils

4.8.1. DETERMINATION OF COEFFICIENTS OF PERMEABILITY OF COHESIONLESS SOILS

The coefficient of permeability of cohesionless soils is usually determined in a constant-head permeameter (Figure 4.16). Whenever possible the test is carried out on undisturbed samples (extracted in thin-walled tubes or special rings) of known volume and density. The sample in the ring is dried prior to testing and the dry density ρ_d of the sand is established. In the case when undisturbed samples are not available the test is carried out on several samples

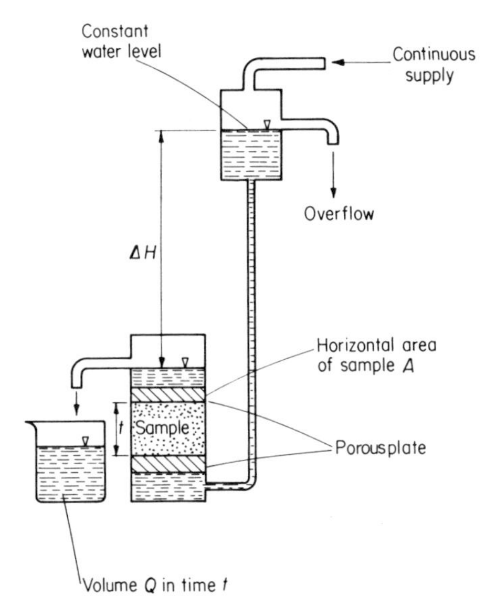

Figure 4.16. Constant-head permeameter.

of the dried sand which is then either loosely poured into the test ring or compacted in it by vibration.

In order to prevent segregation of the sand it is poured into the ring through a funnel held close to the surface of the sand. On filling the ring (with a removable extension) the sand is densified by striking the wall of the ring with a vibrating fork. Vibration is stopped when the required density of the sample has been achieved.

On placing the ring with the dry sand in the apparatus water is slowly allowed to permeate the sample from the bottom (in order to displace all the air) and at least five readings of the quantity of flow Q are taken at certain time intervals t; hydraulic gradient is varied by moving the header tank.

The results are evaluated using Equation (4.7) and the values of the coefficient k are reduced to k_{10} using Equation (4.5) and the recorded temperatures of the water T.

The evaluated values of k_{10} for different degrees of compaction, i.e. for different ρ_d, are presented in the form of a graph as shown in Figure 4.17.

If the tests are carried out for the purpose of lowering the ground water table by means of *well point* depression wells, then the representative coefficient of permeability k_{10} is taken as that corresponding to the *in situ* dry density. In Figure 4.17, for $\rho_{d\,nat} = 1{\cdot}630$ g/ml, $k_{10} = 3{\cdot}4 \times 10^{-2}$ mm/s is obtained.

When the tests are carried out on sands for road construction purposes, then the authors suggest that the minimum value of the coefficient of permeability k_{10} mm should be taken which corresponds to the maximum dry density $\rho_{d\,max}$, i.e. to the condition of the maximum compaction of the sand beneath the road surface. In Figure 4.17 the coefficient $k_{10} = 2 \times 10^{-2}$ mm/s.

When water permeates through a sand there is a tendency for the dissolved air to be liberated in the form of small air bubbles. This will reduce the volume

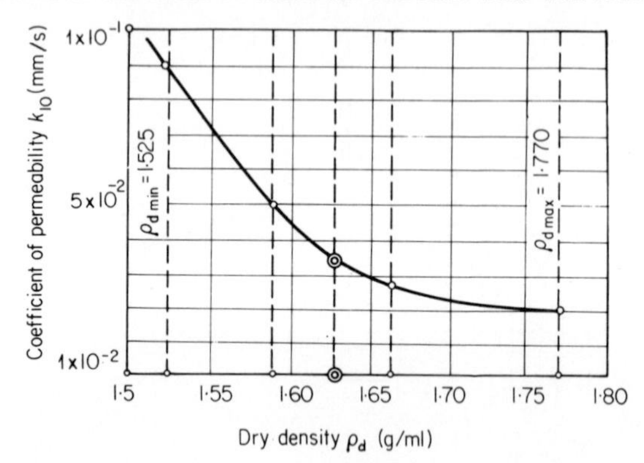

Figure 4.17. Relationship between the coefficient of permeability k_{10} and dry density ρ_d ($\rho_{d\,min} = 1{\cdot}525$ g/ml, $\rho_{d\,nat} = 1{\cdot}630$ g/ml, $\rho_{d\,max} = 1{\cdot}770$ g/ml).

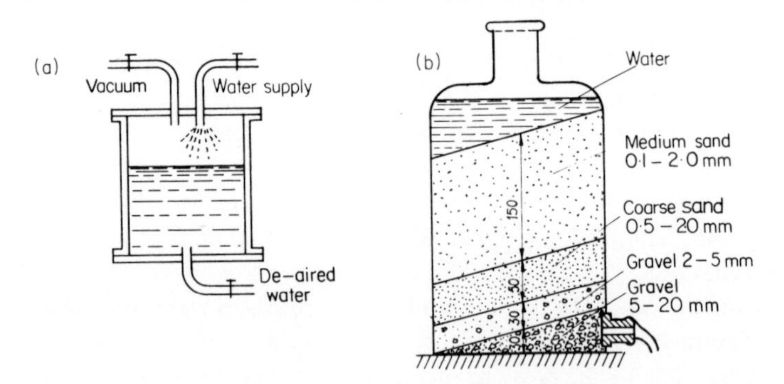

Figure 4.18. De-airing apparatus.

of the pores through which the water can pass and may lead to a considerable underestimate in the value of k_{10}. In order to eliminate errors from the above source the water is de-aired prior to permeability tests either by spraying it into vacuum (Figure 4.18(a)) or by allowing it to stand for a long time and then by passing it through a thick sand gravel filter (Figure 4.18(b))–in both cases the water should be at room temperature.

Example. Data: soil, medium sand;

$$\Delta H = 26 \text{ mm}, \quad l = h = 60 \text{ mm},$$

$$Q = 1000 \text{ ml}, \quad A = 10\,000 \text{ mm}^2, \quad t = 180 \text{ s}, \quad T = +18{\cdot}6\,°\text{C}$$

Determine coefficient of permeability k_{10}.

$$i = \frac{26}{60} = 0{\cdot}43$$

$$k_T = \frac{Q}{iAt} = \frac{1000 \times 10^3}{0{\cdot}43 \times 10\,000 \times 180} = 1{\cdot}29 \approx 1{\cdot}3 \text{ mm/s}$$

$$k_{10} = \frac{1{\cdot}3}{0{\cdot}7 + 0{\cdot}03 \times 18{\cdot}6} \approx 1{\cdot}0 \text{ mm/s}$$

4.8.2. DETERMINATION OF COEFFICIENTS OF PERMEABILITY OF COHESIVE SOILS

The coefficient of permeability of cohesive soils is usually determined in a falling-head permeameter which can be in the form of a specially adapted oedometer (Figure 4.19). In this type of test the hydraulic gradient $i = \Delta H/l$

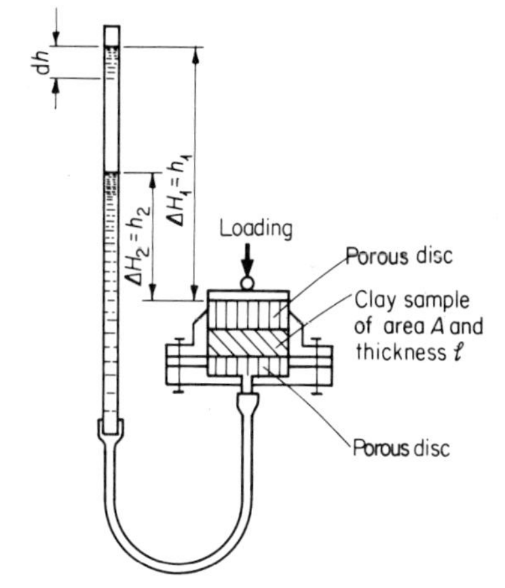

Figure 4.19. Determination of coefficient of permeability of cohesive soil in a 'falling-head' test.

is variable. Since the seepage of water through cohesive soils is very slow, the quantity of flow in a given time interval is observed in a small-diameter glass stand-pipe of cross-sectional area 'a' which is connected with a rubber tube to the oedometer; the drop in levels from h_1 to h_2 in a time interval $t = t_2 - t_1$ is observed.

In time dt the water level in the stand-pipe drops through dh resulting in a loss of water from the system equal to

$$\mathrm{d}Q = -a\,\mathrm{d}h$$

According to Darcy's law the amount of water flowing in time $\mathrm{d}t$, through a soil sample of length l and cross-sectional area A under a head $\Delta H = h$, is

$$\mathrm{d}Q = kA\frac{h}{l}\mathrm{d}t$$

and therefore

$$-a\,\mathrm{d}h = kA\frac{h}{l}\mathrm{d}t$$

$$-\frac{\mathrm{d}h}{h} = \frac{kA}{al}\mathrm{d}t$$

Integrating this equation

$$[-\ln h]_{h_1}^{h_2} = \frac{kA}{al}[t]_{t_1}^{t_2}$$

$$\ln\frac{h_1}{h_2} = \frac{kA(t_2 - t_1)}{al}$$

or

$$k = \frac{al}{A(t_2 - t_1)}\ln\frac{h_1}{h_2} \qquad (4.14)$$

Equation (4.14) enables one to determine the coefficient of permeability of a cohesive soil from a falling-head test. When the permeability is determined in an apparatus in which the sample is contained in a rigid ring (e.g. in an oedometer), the inner surface of the ring should be coated with a hydrophobic material to prevent the flow of water between the sample and the confining ring.

Example. Data: soil, – clayey silty sand;

$$h_1 = 1000\ \text{mm},\quad h_2 = 500\ \text{mm},$$

$$t_1 = 0,\quad t_2 = 6000\ \text{s},\quad a = 100\ \text{mm}^2,\quad A = 3330\ \text{mm}^2,\quad l = 20\ \text{mm},\quad T = 22\,^\circ\text{C}$$

Determine the coefficient of permeability k_{10}.

$$k_T = \frac{al}{A(t_2 - t_1)}\ln\frac{h_1}{h_2} = \frac{100 \times 20}{3330 \times 6000}\ln\frac{1000}{500} \approx 3 \times 10^{-5}\ \text{mm/s}$$

$$k_{10} = \frac{k_T}{0{\cdot}7 + 0{\cdot}03T} = \frac{3 \times 10^{-5}}{0{\cdot}7 + 0{\cdot}03 \times 22} = \frac{3 \times 10^{-5}}{1{\cdot}36} \approx 2{\cdot}2 \times 10^{-5}\ \text{mm/s}$$

Permeability of cohesive soils can also be determined in triaxial compression apparatus of the type shown in Figure 5.39. The test is carried out under a constant head, supplied from one of the self-compensating mercury control systems, and the quantity of flow is measured with a simple burette or with a special volume change indicator to which constant back-pressure can be supplied from another mercury control system (if available). The cell pressure is kept constant throughout the test and must at all times be greater than the water pressure which induces the flow; to minimize the possibility of water flowing between the sample and the rubber membrane the inside of the membrane should be coated with a hydrophobic material (e.g. silicon grease). The coefficient of permeability k_{10} is evaluated using Equations (4.7) and (4.5).

It is necessary to ensure that the test is carried out under hydraulic gradients 'i' greater than the initial gradients i_0 (Roza, 1950). In cohesive soils

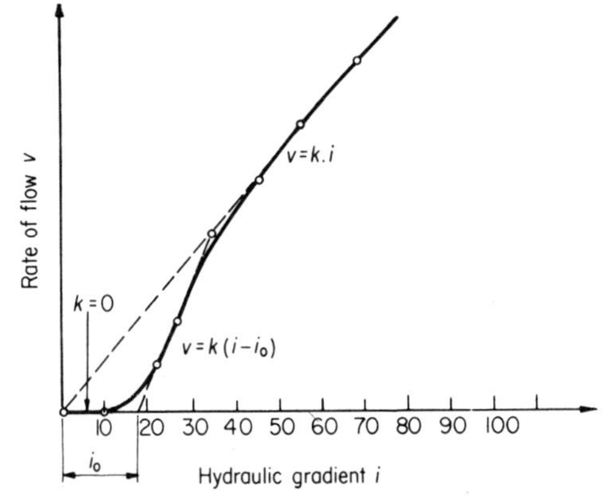

Figure 4.20. The initial hydraulic gradient in cohesive soils.

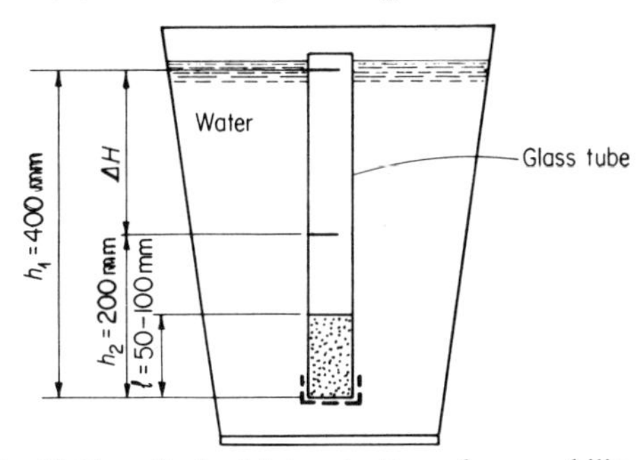

Figure 4.21. Field method of determination of permeability of sands.

the pores are almost completely filled with the adsorbed water and under hydraulic gradients smaller than i_0 there is no flow of water through the soil (Figure 4.20). According to Roza the initial hydraulic gradients for cohesive soils can be greater than 10.

The relationship between the rate of flow of water and the hydraulic gradient in cohesive soils, clays in particular, can be illustrated as shown in Figure 4.20.

The above-described falling-head test can also be used for a rapid field determination of the permeability of cohesionless soils. Simple apparatus, as illustrated in Figure 4.21, can easily be assembled in the field. A glass tube of diameter between 40 and 60 mm and of length between 450 and 500 mm has divisions marked on it at distances of 400, 200, 100, and 50 mm from the lower end; the lower end is covered with a wire gauze 0·5 to 1 mm or with ordinary cotton gauze.

The sand is deposited in the tube (through a funnel with a rubber tube extension) to a height of 50 to 100 mm and is densified by striking the tube with a wooden fork. It is then slowly immersed in water which has been boiled (to de-air it) and cooled down to the surrounding temperature. The tube is immersed until the water level inside it is well above the 400 mm mark. It is then completely withdrawn from the water in the bucket and when the water level in the tube reaches the 400 mm mark a stop-watch is started. The stop-watch is stopped when the level in the tube reaches either the 200 or 100 mm mark.

The coefficient of permeability is then determined from the following equation:

$$k_{10} = \frac{l \times e}{t(0{\cdot}7 + 0{\cdot}03T)} \tag{4.15}$$

where l = height of sand sample in mm

t = time taken for water to drop from level h_1 to h_2 in seconds

$e = \ln \frac{h_1}{h_2}$; for h_1 = 400 mm and h_2 = 200 mm, e = 0·693; and, for h_2 = 100 mm, e = 1·386

T = temperature in water in °C

The determination of the coefficient of permeability k_{10} is particularly important in the case of sands used in road construction as a sub-base material or for land drains; in the maximum state of compaction I_D = 1·0 such sands should have the following values of k_{10}:

(a) sands for sub-base $k_{10} \geqslant 5$ mm/min

(b) sands for land drains $k_{10} \geqslant 10$ mm/min

4.9. Laboratory Methods of Determination of Pore Water Suction Forces in Soils

The problem of the capillary rise of water above the phreatic surface (water table) and of the movement of water in homogeneous soils from higher to lower water content zones is related to the phenomenon of pore water suction in soils (see Sections 2.5 and 2.6). This problem is of great importance in road construction, where the upper formation layers can be subject to periodic drying or wetting.

Of particular importance is the problem of freezing of soils which frequently leads to the reduction in their strength in spring time and failure of the road surfacing. These problems are also of considerable importance in foundation engineering.

The suction forces are determined from the measurement of pore water suctions in soils during (1) wetting process, and (2) extraction of water from the soil.

The determination of suction in a soil during the wetting process can be carried in an apparatus shown schematically in Figure 4.22. On filling the apparatus with de-aired water the soil sample (contained in a ring) is placed

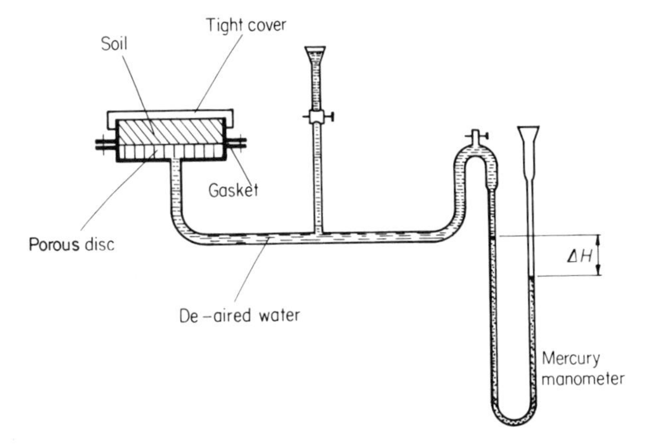

Figure 4.22. Apparatus for measurement of suction forces during wetting of soil.

and tightly bedded on the porous base. On achieving equilibrium of the pore water suction throughout the sample and the apparatus, the sample is removed from the ring and its water content is determined. The magnitude of the suction is obtained from the difference of levels of mercury in the manometer.

The investigation is carried out using several samples of different initial water contents. The measured suctions are plotted against the final water contents determined on completion of each test.

The determination of suction during extraction of water from the soil can be carried out in an apparatus shown schematically in Figure 4.23. This apparatus is constructed on the principle of Beskow's capillarometer.

The investigation is also carried out using several samples of different initial water contents. The level of mercury in the open arm of the manometer is slightly raised to push some water through the porous base on which the sample of soil compacted at a predetermined initial water content is placed. The sample container is tightly clamped to the base. The open end of the manometer is then slowly lowered, stopping at intervals for a few minutes, until the air enters through the sample. At that moment the level of mercury in the open arm is noted. The magnitude of suction p is then determined from the following expression:

$$p = (H_0 - H_l)\gamma_w + (H_l - H_r)\gamma_m \tag{4.16}$$

where H_0 = level of the bottom of the sample in m
H_l = level of mercury in the left (fixed) arm in m
H_r = minimum level of mercury in the right (open and moveable) arm in m
γ_w = unit weight of water in kN/m^3
γ_m = unit weight of mercury in kN/m^3

At the same time the final water content of the sample is determined.

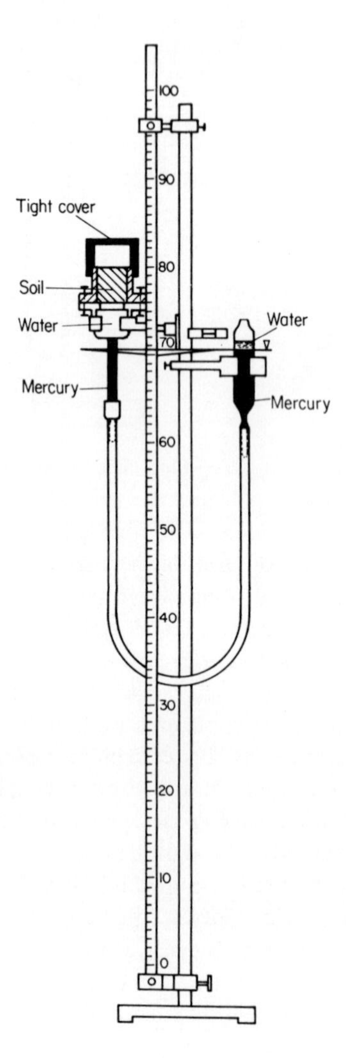

Figure 4.23. Apparatus for measurement of suction forces during extraction of water from soils.

Example. Data as follows:

$$H_0 = 0{\cdot}612 \text{ m}, \quad H_1 = 0{\cdot}536 \text{ m}, \quad H_r = 0{\cdot}323 \text{ m}$$

The pore water suction

$$p = (0{\cdot}612 - 0{\cdot}536)\,9{\cdot}81 + (0{\cdot}536 - 0{\cdot}325)\,13{\cdot}55 \times 9{\cdot}81$$
$$= 0{\cdot}75 + 28{\cdot}06 = 28{\cdot}81 \text{ kN/m}^2$$

The results of the test can be plotted to give the relationship between the pore water suctions and the final water contents of the soil as shown in Figure 4.24.

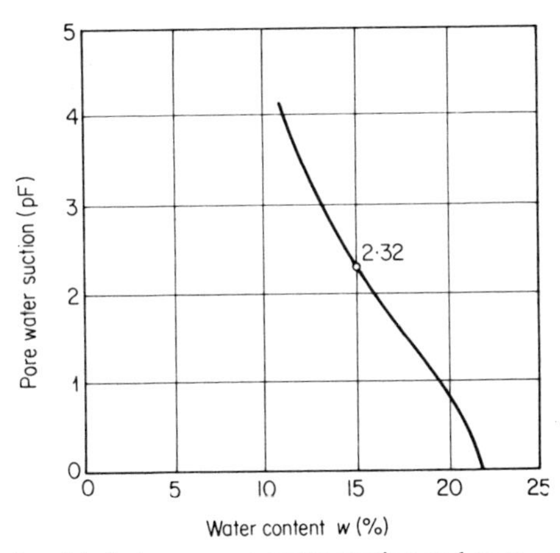

Figure 4.24. Relationship between pore water suction and water content of the soil.

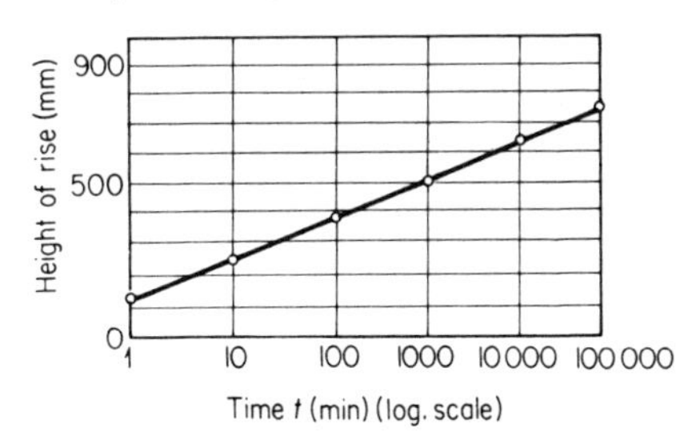

Figure 4.25. The rate of rise of water in soil.

Both the described methods of measurement of pore water suction are not very accurate. In the first apparatus the quantity of water present may not be sufficient to achieve the full equalization of pressures and therefore it is suggested that the manometer tube should have as small a bore as possible and that its free end should be sealed. In the second apparatus the measured suctions are governed by the largest pores and may not be representative of the *in situ* conditions.

The maximum suction that can be measured in the above instruments is of the order of 0·8 atmosphere (8 m head of water) which is usually sufficient for investigations associated with road construction problems (investigation of slightly and medium cohesive frost susceptible soils). In the case when more accurate results are required, involving measurements of higher suctions, other methods are available which involve the use of more sophisticated apparatus (Croney and Coleman, 1961) or special oedometer tests are carried out (Sections 2.6.3 and 5.2). In the oedometer test, several samples of the same soil are tested under different loads. After equilibrium has been reached the water content of the soil is determined. The results are presented in the form of a graph relating the effective stresses and equilibrium water contents.

The above investigations are usually carried out only in the case of cohesive soils. For cohesionless soils the equilibrium pore water suction is only slightly dependent on the water content and there is practically no difference between the wetting and extraction of water processes; therefore field methods of measurement can be used. A simple field test can be carried out using a 0·5 m long glass tube of 40–50 mm diameter which is filled with air-dried sand: the sand is retained in the tube by covering its bottom end with pervious fabric. To avoid segregation it is recommended that the sand is poured through a funnel with a rubber tube extension. The sand is then densified by tapping the walls of the glass tube with a wooden fork.

The tube, filled with sand, is then placed in a beaker into which water is poured. As the water rises owing to the capillary action the colour of the sand changes to a deeper hue.

The level of the capillary rise is recorded at 1, 2, 5, 10, 15, and 30 minutes and 1, 2, 4, 6, 24, and, possibly, 48, 72, etc., hours after the start of the test. It is assumed that the equilibrium condition has been reached when the level remains stationary for 24 consecutive hours.

The results are plotted on semilogarithmic graph paper as shown in Figure 4.25; the height of the capillary rise can be taken as the intersection of the produced experimental line with the vertical line through the point $t = 100\,000$ min. The height of the capillary rise is equal to the pore water suction of the given soil.

4.10. Movement of Capillary Water in Soils

The capillary water moves from a zone of lower to a zone of higher pore water suction. The movement of the water in the vertical direction is subject to the action of the gravitational force; in addition, it is resisted by the presence of the adsorbed water. Therefore the movement of water from a lower to a higher level will not take place unless the difference in the suctions (expressed in terms of head of water in m) is greater than the vertical distance between the two levels and the resistance of the adsorbed water. This explains the fact why even under tropical conditions clays can be quite wet at greater depths while their surface is quite dry.

If the gravitational force did not exist and there was no resistance by the adsorbed water, then in a homogeneous soil the water content would be constant throughout its entire formation; any local change in it would result in the movement of water over the entire formation leading to the establishment of equilibrium at another constant water content.

In the case of layered or laminated soils, having different equilibrium pore water suctions, the equalization process leads to considerable increases in the water content of the more clayey soils and to decrease in the water content of the less clayey soils, e.g. in varved clays the clay laminae have higher water content than the silt laminae.

If the upper horizons of the soil are dryer than the lower (owing to desiccation of the surface) then, on covering it by, for example, a road surface, equalization of water contents will tend to take place in the soil and may lead to a decrease in its bearing capacity. Therefore the determination of bearing capacity and compressibility of a soil on the basis of plate-bearing tests carried out during drought periods is dangerous and should be avoided.

During frost periods the formation of ice lenses in the frozen zone of the soil leads to a decrease in the water content of the underlying formation thus increasing its pore water suction; this in turn leads to an upward movement of water from deeper and wetter formations. The net result is an increase in the water content of the frozen zone; the high water content in this zone leads to failure of road pavements in the spring time.

All the above cases of water content changes in soils are of considerable importance in road engineering. In each case the changes are induced by one or several of such occurrences as a change in external loading, lowering, or rising of the ground water table and desiccation or wetting of soil, etc.; they involve a change from the initial state of equilibrium of the pore water suction to a new state that corresponds to the new conditions.

For road construction purposes it is important to be able to predict the changes in the state of equilibrium of water contents at different depths in the soil.

The analysis involves the use of Equation (2.9) and of a relationship between the effective stresses in the soil and its water content.

4.11. Rate of Flow of Capillary Water

If there is movement of water due to the differences in pore water suctions, then the rate of flow can be computed (without considering the action of gravitational force) from an equation similar to Darcy's law:

$$v_c = k_p\left(\frac{p_1 - p_2}{l} - i_0\right) \tag{4.17}$$

where p_1 and p_2 = pore water suctions at two different levels
l = distance between the levels
k_p = seepage coefficient dependent on the magnitude of the pore water suction
i_0 = initial hydraulic gradient necessary to overcome adsorbed water resistance

According to the results of the tests carried out in U.S.A. the value of k_p for slightly cohesive soils is of the order of 3×10^{-5} m/day at the pore water suction of 0·2 kgf/cm^2 (approximately 20 kN/m^2) and of the order of 15×10^{-5} m/day at the pore water suction of 0·07 kgf/cm^2 (approximately 7 kN/m^2). The above values have been obtained from Equation (4.17) without considering i_0.

5

Mechanical Properties of Soils and Laboratory Methods of their Determination

5.1. Deformation of Soils

A change in a system or intensity of loading acting on a given medium will result in its deformation. Solid media of a continuous atomic structure (e.g. steel) exhibit relatively small and almost instantaneous deformations, while deformations of granular media are generally large and take place over a long period of time; the rate of deformation decreases with time until equilibrium is reached between the external and internal forces.

The essentially granular character of all soils therefore ensures that they experience large and prolonged deformations over a period of time depending on porosity, permeability, and the nature of the interparticle bonds. These specific soil characteristics necessitate the use of special testing techniques for determination of their stress–strain relationship and special methods of computation of deformations.

The characteristics which govern the magnitude and the rate of deformation of soil media are known as mechanical properties and the two most important, compressibility and shear strength, will now be discussed in detail.

5.2. Compressibility of Soils

The compressibility of a given medium is defined as its capacity to decrease in volume on application of loading.

In foundation engineering one is mainly concerned with vertical movements and it is usual to express the compressibility of soils in terms of a relationship between vertical stresses and strains. If a soil exhibited linear elasticity, then one would not hesitate to use the modulus of elasticity E as a measure of its compressibility.* Soil, however, is not an elastic medium, its stress–strain relationship being almost invariably non-linear, and it undergoes permanent

* Modulus of elasticity is a reciprocal measure of compressibility, i.e. a high value of E implies low compressibility and vice versa.

deformations, i.e. on unloading it does not recover to its original volume and the compression curve 1 (Figure 5.1) does not coincide with the unloading curve 2.

Taking the above points into consideration, the compressibility of soil can still be defined by a quasi-elastic secant modulus E, assuming that over a small

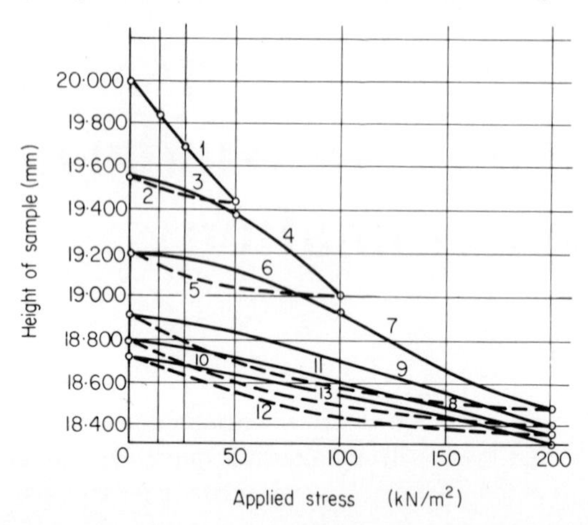

Figure 5.1. Typical compression and expansion curves: 1, 4, and 7, first loading compression curves; 2, 5, 8, 10, and 12, expansion or unloading curves; 3, 6, 9, 11, and 13, reloading curves (second compression curves).

range of stresses the stress–strain relationship is approximately linear. For any given soil the value of E will not only be stress range dependent but will also be different for the first loading, unloading, and subsequent reloading: it will also be dependent on whether the sample was tested in a laterally unconfined, confined, or partly confined manner (Sections 5.2.5 and 5.2.7). A quasi-elastic secant modulus, determined from a specific axial compression test in which the soil sample is laterally confined (e.g. the oedometer test) and hence in which the axial strain is equal to the volumetric strain, will be used as a practical measure of the compressibility of soils and will be termed a stiffness modulus E.*

A decrease in the volume of a soil specimen on application of loading is equal to the decrease in the volume of its pores (voids) resulting from relative movement of individual particles with changes in thickness of the adsorbed water films at the points of contact (Figure 5.2). Changes in volume of the particles themselves are so small as to be negligible.

If the soil is cohesive and fully saturated, then initially practically the entire load increment will be supported by the pore water; this will be followed by expulsion of pore water from highly stressed zones to those of lower stress

* The above-defined stiffness modulus is equal to reciprocal of Terzaghi's coefficient of volume compressibility m_v.

intensity. In the case of soils of low permeability (cohesive soils) the expulsion of water from the pores will take a long time. Deformation of these soils is therefore much slower than of the more permeable cohesionless soils, whose deformation is almost complete immediately on application of the loading.

If soils are not fully saturated, then the liquid phase contains small bubbles of air or gas which compress virtually instantaneously with an increase of the pore water pressure. Under these circumstances even cohesive soils will exhibit some immediate deformation on application of loading while the rest of the deformation will then take place at a slow rate governed by the expulsion of the pore water from the highly stressed zones of the soil; this will be accompanied by dissipation of the pore water pressure and an increase in the

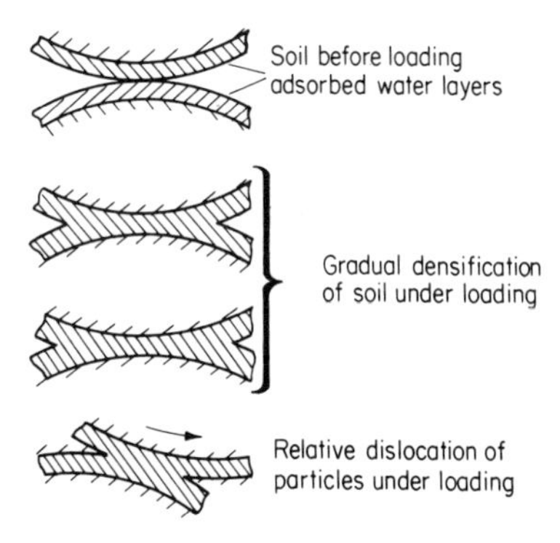

Figure 5.2. Deformation of adsorbed water film at the point of contact between two particles (after Terzaghi).

volume of the air bubbles. The expulsion of water from the pores continues until the pore water pressure in the stressed zone equalizes with the pore water pressure in the zones outside the influence of the loading.

In overconsolidated soils in which the pores are almost completely filled with adsorbed water the pore water pressures may remain higher within the stressed zone than on the boundaries if the induced hydraulic gradients are smaller or equal to the initial hydraulic gradient (Section 4.8.2).

On removal of external loading the soil recovers and increases in volume because the resulting decrease in the effective interparticle stresses allows the increase in thickness of the adsorbed water films at the points of contact. Obviously the soil does not recover to its initial volume because on loading some of the particles would have become permanently bonded together and some would have been displaced (Figure 5.2).

After each loading and unloading the soil deforms partly elastically and partly permanently. Soil which has been subjected to several loading cycles begins to acquire the characteristics of an elastic body (Figure 5.1, curves 9, 11, and 13) in which case it is possible to talk about certain elasticity of soils.

Within the limited working stress ranges which are possible in soils, it is reasonable to assume that soils behave as linearly deforming media and that it is possible to apply to them the principles and equations that have been developed for elastic media.

5.2.1. DETERMINATION OF COMPRESSIBILITY OF SOILS

The investigation of compressibility and expansion characteristics of soils is usually carried out in oedometers (Figure 5.3).

Figure 5.3. Typical mechanical oedometer (by courtesy of Wykeham Farrance Engineering Ltd.) and hydraulic oedometer (by courtesy of Armfield Engineering Ltd.).

The oedometer test is basically a model test in which deformations of a soil sample laterally confined in a rigid metal ring are observed under direct incremental loading (Figure 5.4). This, to a large extent, corresponds to the conditions under which the same soil element would be stressed below a foundation, where the lateral strains are partly prevented by the neighbouring soil elements. The stiffness modulus obtained from an oedometer test will be referred to as the oedometric modulus E_{oed}.

Cylindrical samples are used having height/diameter ratios between 0·25 and 0·33 (British Standard 1377, 1967). During the test, the soil is prevented from drying by placing a rubber membrane cover over the cell or by surrounding the sample with water. In the latter case it is necessary to apply an initial load increment to prevent the sample from swelling. The magnitude of the initial load increment (also known as the equilibrium stress) mainly depends on the

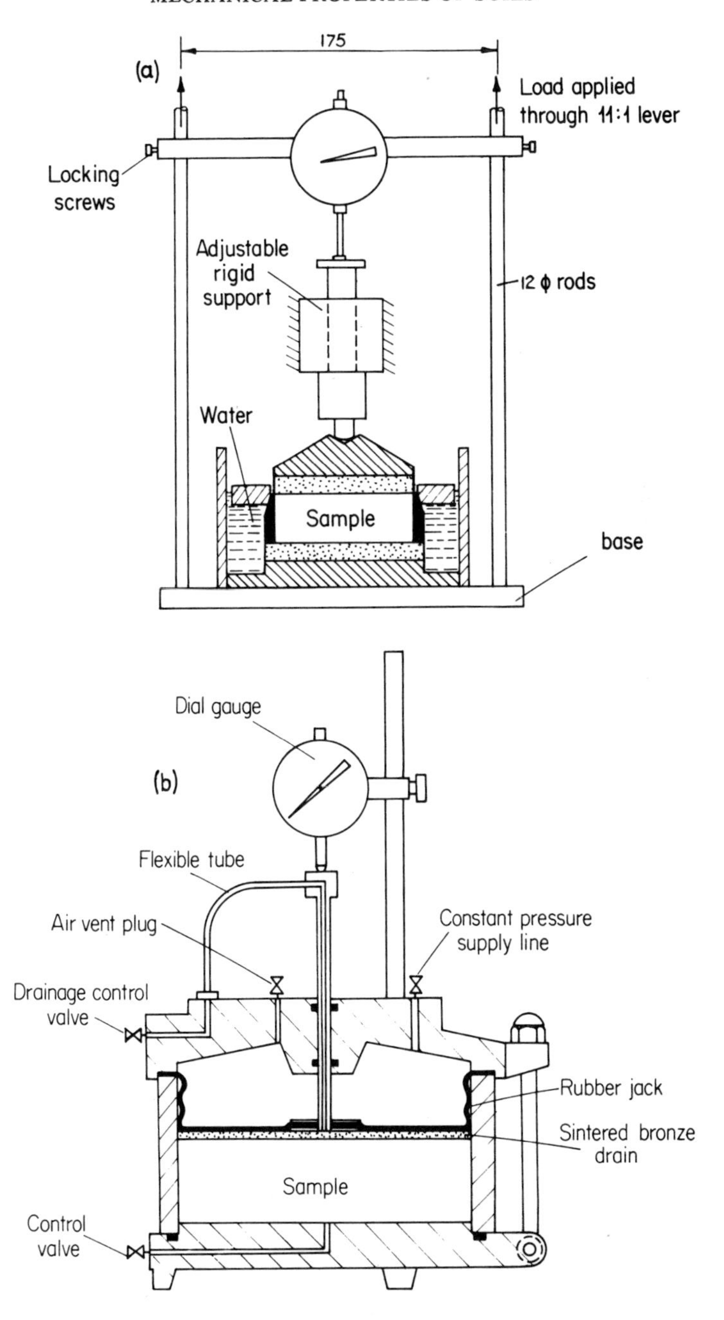

Figure 5.4. Schematic details: (a) typical mechanical oedometer cell; (b) hydraulic oedometer.

effective overburden stress but can also be related to the consistency of the soil under consideration: for stiff soils it should be taken as equal to the estimated effective overburden stress, for firm soils as somewhat less and for soft soils as appreciably less than the effective overburden stress. For very stiff and very hard soils the initial load increment is usually greater than the effective overburden stress and its determination should form part of the test (Section 5.2.7). Metastable soils such as loess require special treatment (Section 5.2.6).

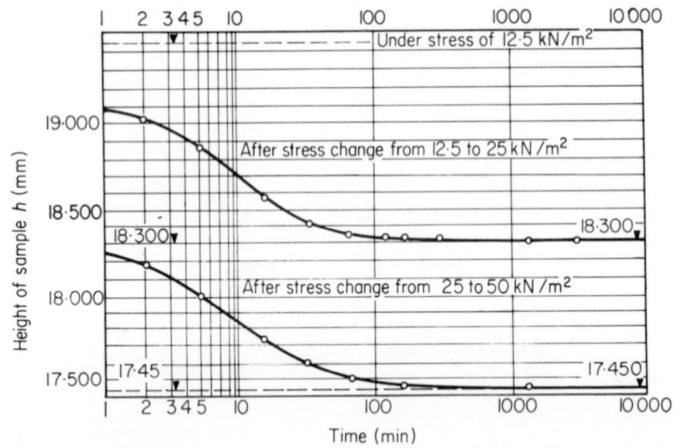

Figure 5.5. Consolidation curves for consecutive 12·5 and 25 kN/m² load increments.

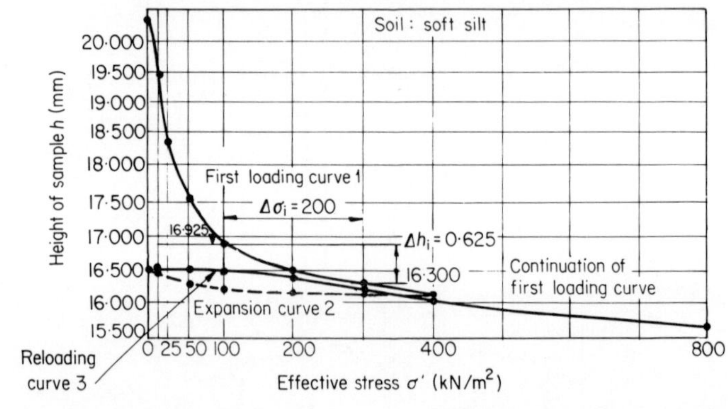

Figure 5.6. Compression curves for first loading, unloading (expansion), and reloading.

On completion of the initial stage of the test the sample is loaded by increasing the intensity of loading to twice its previous value, e.g. 12·5, 25·0, 50 kN/m², etc. (250, 500, 1000 lbf/ft² or 0·125, 0·25, 0·5 kgf/cm²) until the required stress range has been covered or at least four load increments have been added; in unloading (expansion) test the loading intensity is halved at each stage.

On each application or removal of a load increment changes in thickness of the sample are recorded after 0·5, 1, 2, 5, 10, 15, and 30 minutes and also after

1, 2, 4, 8, and 24 hours. Typical consolidation curves plotted in semilogarithmic scale are shown in Figure 5.5; the initial parts of the curves (up to approximately 500 min) illustrate the process of the primary consolidation, which is governed by expulsion of the pore water, while the linear tails illustrate the process of the secondary consolidation, governed by deformation of the adsorbed water films at the points of contact. The secondary effects are usually small in the case of insensitive inorganic soils but can be predominant in sensitive or organic and peaty soils.

From total changes in thickness during each incremental loading stage a complete compression or expansion curve is built up as shown in Figure 5.6.

5.2.2. FIRST LOADING AND RELOADING STIFFNESS MODULI

When a remoulded soil (at a water content close to its liquid limit) is consolidated in an oedometer then the so called first compression curve 1 is obtained (Figure 5.6). If on reaching a certain stress the soil is unloaded, then an unloading (expansion) curve 2 is obtained which lies well below the first compression curve. On reloading, the trace follows curve 3 which initially lies above the unloading curve but then crosses it, forming a hysteresis loop, and gradually becomes the continuation of the first compression curve.

Curve 1 is known as the first or virgin compression curve and curve 3 as the reloading compression curve. As can be seen the slope of curve 3 is much smaller than the slope of curve 1 which indicates that the soil on reloading is less compressible than on the first loading.

The oedometric first loading modulus in the stress range between σ_i and σ_{i+1} is obtained from the following expression:

$$E'_{\text{oed}} = \frac{\Delta\sigma_i}{\Delta h_i/h_i} = \Delta\sigma_i h_i/\Delta h_i \qquad (5.1)$$

where E'_{oed} = first loading oedometric modulus (for no lateral strain condition), in the same units as $\Delta\sigma_i$

$\Delta\sigma_i$ = stress increment, $\Delta\sigma_i = \sigma_{i+1} - \sigma_i$ (Figure 5.6), in kN/m^2 (lbf/ft^2, kgf/cm^2)

Δh_i = change in thickness of sample due to stress increment $\Delta\sigma_i$, in mm (in)

h_i = thickness of soil sample prior to application of stress increment $\Delta\sigma_i$, in mm (in)

$\Delta h_i/h_i$ = vertical strain corresponding to stress increment $\Delta\sigma_i$

The value of the oedometric modulus obtained from Equation (5.1) must be increased to compensate for two sources of error inherent in sampling processes and testing apparatus: (1) disturbance to the soil structure during sampling (Hvorslev, 1949a), subsequent extrusion from the sampler and transfer to the oedometer ring; (2) bedding in, due to lack of fit between the soil sample, the confining ring, and the top and bottom porous stones and also due to slackness within the oedometer itself (Wiłun, 1955a); mechanical oedometers should be calibrated by testing in them materials of known compressibility, e.g. rubber. The magnitude of the correction coefficient κ depends upon the value of E_{oed} and can be taken from Table 5.1 (Wiłun, 1955a); then

$$E' = \kappa_f E'_{oed} \tag{5.1a}$$

and

$$E'' = \kappa_r E''_{oed} \tag{5.1b}$$

where E' and E'' are referred to as the first and reloading stiffness moduli.

A correction coefficient κ greater than 2 indicates the presence of unacceptable errors and hence the unsuitability of the oedometer test for soils of low compressibility; it is suggested by the authors that the use of the oedometer test should be limited to cohesive soils of plastic consistency (I_c between 0·0 and 0·75) and to organic silty clays (recent alluvium).

Table 5.1. Correction coefficient κ for determination of first loading and reloading stiffness moduli

Uncorrected oedometric modulus E_{oed} (MN/m²)	Correction coefficient	
	For first loading κ_f	For reloading κ_r
0·2*	1·00	1·00
0·5	1·02	1·01
0·7	1·03	1·02
3·0	1·10	1·03
5·0	1·25	1·06
7·0	1·50	1·10
10·0	2·00	1·17
15·0	(3·00)	1·35
20·0	(4·00)	1·60
25·0	(6·00)	2·00
30·0 and over	(10·00)	(3·00)

* For approximate conversion to tonf/ft² and kgf/cm² multiply by 10.

Equation (5.1) is taken to be in accordance with Hooke's law on the condition that its use must be limited to small stress ranges; this is generally acceptable because for small load increments soil deformations are almost linearly related to the stress increases.

Example. Determine the first loading stiffness modulus E' from the first compression curve (Figure 5.6) in stress range 100 to 300 kN/m².

$$\Delta\sigma_i = 300 - 100 = 200 \text{ kN/m}^2$$
$$\Delta h_i = 16{\cdot}925 - 16{\cdot}300 = 0{\cdot}63 \text{ mm}$$
$$h_i = 16{\cdot}93 \text{ mm}$$
$$E'_{oed} = (200 \times 16{\cdot}93)/0{\cdot}63 \approx 5400 \text{ kN/m}^2 = 5{\cdot}4 \text{ MN/m}^2$$

From Table 5.1, for $E'_{oed} = 5{\cdot}4$ MN/m² the correction coefficient $\kappa_f = 1{\cdot}3$. Therefore the corrected oedometric stiffness modulus, i.e. the stiffness modulus $E' = 5{\cdot}4 \times 1{\cdot}3 \approx 7{\cdot}0$ MN/m².

For determination of reloading or expansion stiffness modulus the correction factor κ_r (Table 5.1) is used.

$$E'' = \kappa_r E''_{oed}$$

Example. Determine the reloading stiffness modulus E'' from the reloading curve (Figure 5.6) in the stress range 100 to 300 kN/m².

$$\Delta\sigma_i = 200 \text{ kN/m}^2$$
$$\Delta h_i = 16{\cdot}40 - 16{\cdot}15 = 0{\cdot}25 \text{ mm}$$
$$h_i = 16{\cdot}40 \text{ mm}$$
$$E''_{oed} = (200 \times 16{\cdot}40)/0{\cdot}25 \times 10^3 \approx 13{\cdot}0 \text{ MN/m}^2$$

hence
$$E'' = 1{\cdot}28 \times 13{\cdot}0 \approx 17{\cdot}0 \text{ MN/m}^2$$

It can be seen from the above examples that soils which previously have been substantially preconsolidated, e.g. by a glacier or soil overburden, have considerably higher stiffness moduli than their normally consolidated post-glacial counterparts.

Average values of the elastic modulus E for rocks and stiffness moduli E' and E'' for soils are given in Table 5.2.

It must be emphasized that laboratory methods for determining the compressibility of soils are on the whole inaccurate and can only be used with confidence in the case of soils of high compressibility. In the case of soils of low compressibility the values of E obtained from oedometer tests are usually too low. Therefore it is recommended that in the design of large and important structures *in situ* loading tests should be carried out from which more accurate values of stiffness moduli can be obtained. The details of *in situ* loading tests and the interpretation of their results are given in Chapter 6.

5.2.3. COEFFICIENT OF COMPRESSIBILITY AND COEFFICIENT OF VOLUME COMPRESSIBILITY

It is sometimes convenient to express the compressibility of soils in terms of coefficient of compressibility a_v. This parameter relates the change in the voids ratio of a given soil under conditions of zero lateral strain with the change in the effective stress:

$$a_v = \Delta e_i/\Delta\sigma_i \tag{5.2}$$

where a_v = coefficient of compressibility
Δe_i = change in voids ratio due to change in stress of $\Delta\sigma_i$
$\Delta\sigma_i = \sigma_{i+1} - \sigma_i$

A relationship can be developed between the coefficient of compressibility and the oedometric modulus. By considering an element of soil having a unit base area $A = 1$ and containing a unit volume of solids $V_s = 1$ then the volume of voids in this element is equal to the voids ratio e_i (Figure 5.7(a)). The height of such an element can be taken as $h_i = 1 + e_i$. On application of loading the volume of the voids in the element will decrease by $\Delta e_i = e_i - e_{i+1}$. If it is assumed that because the compressibility of the soil particles is very low

Table 5.2. Average values of elastic and stiffness moduli (after Polish Code PN-59/B-03020, 1959)

Description of rocks or soils: Class	Group	Type		Elastic modulus E for rocks and stiffness moduli E' and E'' for soils (MN/m^2)*				Remarks
rocks	hard	igneous and metamorphic		massive	lightly jointed	heavily jointed		(1) given values of elastic modulus E and stiffness moduli E' and E'' refer to *in situ* conditions
			E	>10 000	10 000–5000	to be determined from *in situ* loading tests		
	soft	sedimentary: lime-stones, sandstones, and shales	E	>4000	4000–2000			
		chalky marls, clay shales, weakly cemented sandstones, etc.	E	200	100			
natural inorganic soils	stony soils	head, rock debris, etc.	E	to be determined from *in situ* loading tests				(2) interpolate linearly to obtain E' or E'' for intermediate values of density index (I_D) or consistency index (I_c)
	coarse-grained soils	gravels, tills, hoggins, residual soils		voids between coarse grains filled with				(3) E' is used for first loading of post-glacial or more recent soils; E'' is used for first loading of overconsolidated soils and for reloading of the more recent soils
				cohesionless soils	cohesive soils			
			I_D =	1·0 — 0·33	I_c = 1·0	0·50	0·0	
			E'' E'	300–100 200–80	150 100	150–40 100–20	40–5 20–5	
	fine-grained soils	cohesionless soils $I_p \leqslant 1{\cdot}0$		dense	medium	loose		(4) it is assumed that stiffness modulus of cohesionless soils is independent of the degree of saturation
			I_D =	1·0 — 0·67	0·33	0		
		coarse and medium sands	E'' E'	200–150 150–80	150–80 80–50	80–20 50–20		
		fine and silty sands	E'' E'	150–100 100–50	100–50 50–30	50–15 30–15		

		cohesive soils $I_p > 1{\cdot}0$		very hard	very stiff or hard	stiff	firm	soft to very soft	(5) for cohesive soils within the range of frost penetration decrease the values of stiffness moduli by 50%
				$w = w_s$	$I_c = 1{\cdot}0$	$0{\cdot}75$	$0{\cdot}50$	$0{\cdot}0$	
		sandy silts, silts, clayey soils, sand–clays, silt–clays, and clays	E'' E'	120 60	120–60 60–30	60–30 30–12	30–12 12–4	12–2 4–0·2	
		metastable soils–loess		if $I_m \leqslant 0{\cdot}02$ values of E taken as for cohesive soil of the same consistency index I_c if $I_m > 0{\cdot}02$ soil structure is unstable and stiffness modulus is determined from laboratory or *in situ* tests					
		recent alluvial clayey silts (mud)	E'	to be determined from laboratory or *in situ* tests					
natural inorganic soils	soils with traces of organic matter	organic sands	E'' E'	$\frac{30}{15}$ (for $I_D = 1{\cdot}0$); $\frac{5{\cdot}0}{2{\cdot}0}$ (for $I_D = 0{\cdot}33$)					(6) interpolate linearly to obtain E for intermediate values of I_D, I_c, or water content w *(7) for approximate conversion to tonf/ft^2 and kgf/cm^2 multiply by 10
		organic silts	E'' E'	$\frac{20}{10}$ (for $I_c = 1{\cdot}0$); $\frac{2{\cdot}0}{1{\cdot}0}$ (for $I_c = 0{\cdot}5$)					
	organic muds	organic clayey silts (mud)	E'' E'	$\frac{10}{5}$ (for $I_c = 1{\cdot}0$); $\frac{1{\cdot}0}{0{\cdot}5}$ (for $I_c = 0{\cdot}5$)					
	peat	peaty soils and peats	E'' E'	$\frac{5{\cdot}0}{2{\cdot}0}$ (for old peats with $w \leqslant 100\%$;) $\frac{0{\cdot}5}{0{\cdot}2}$ (for recent peats with $w \geqslant 300\%$)					
fills		sandy fills		as for sands–depending on I_D					
		inorganic cohesive fills		to be determined from *in situ* tests					
		organic soil fills		not suitable for founding on					

their volume remains unaltered, the change in the height of the element under consideration will be equal to the change in the voids ratio (Figure 5.7(b)), i.e. $\Delta h_i = \Delta e_i$. Substituting into Equation (5.1)

$$E_i = \Delta\sigma_i h_i/\Delta h_i = \Delta\sigma_i(1 + e_i)/\Delta e_i = (1 + e_i)/a_v$$

If the coefficient of compressibility is required, then the compression curves obtained from oedometer tests are plotted in terms of voids ratio e and the vertical stress σ' (Figure 5.8).

The following equation gives the relationship between e_i and h_i:

$$e_{i+1} = e_i - \Delta h_i(1 + e_i)/h_i \tag{5.3}$$

In his original development of one-dimensional consolidation theory Terzaghi (1943) has introduced a coefficient of volume compressibility m_v which is now

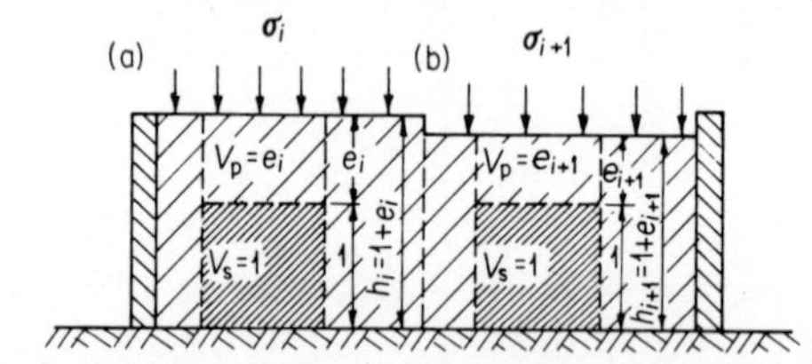

Figure 5.7. Deformation of soil sample in oedometer and change in voids ratio.

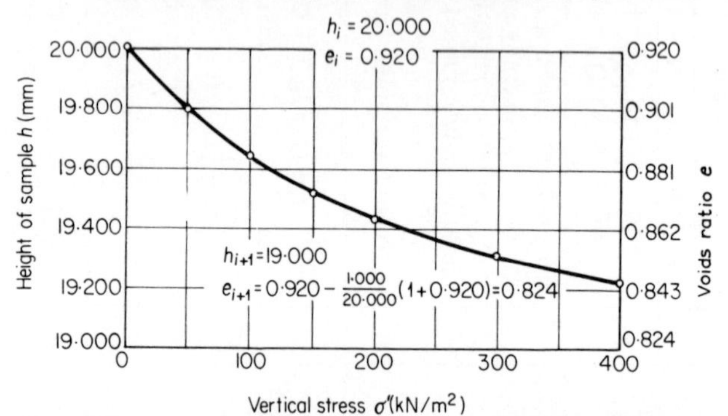

Figure 5.8. Oedometer test results plotted in terms of voids ratio and vertical stress.

widely used as a measure of the compressibility of soils. There is a direct relationship between the coefficient of compressibility, the oedometric modulus, and the coefficient of volume compressibility:

$$E_{oed} = (1 + e_i)/a_{vi} = 1/m_v \tag{5.4}$$

For practical application the values of a_v and m_v should be corrected in the same manner as the oedometric moduli.

5.2.4. COMPRESSIBILITY INDEX

According to Terzaghi the virgin compression curves for remoulded cohesive soils of plastic consistency (between plastic and liquid limits) are of the logarithmic type (Figure 5.9) and for a change in stress $\Delta\sigma_i = \sigma_{i+1} - \sigma_i$ can be defined by an empirical relationship containing the compressibility index C_c:

$$e_{i+1} = e_i - C_c \log \frac{\sigma_i + \Delta\sigma_i}{\sigma_i} \tag{5.5}$$

Similarly the unloading (expansion) curves when plotted in a semilogarithmic scale also appear as straight lines; expansion index C_e is defined as the slope of the unloading lines.

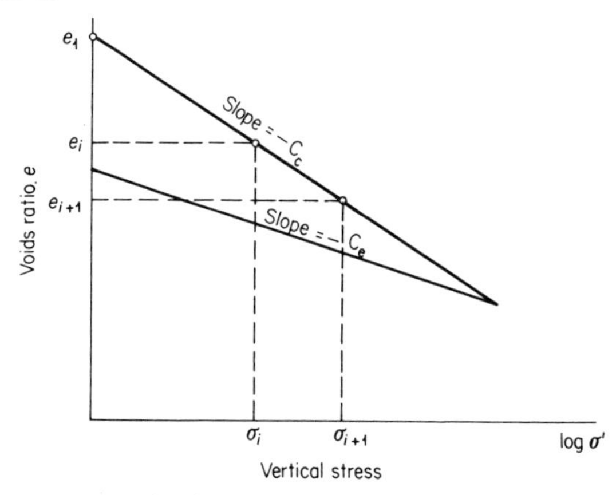

Figure 5.9. Determination of compressibility and expansion indices.

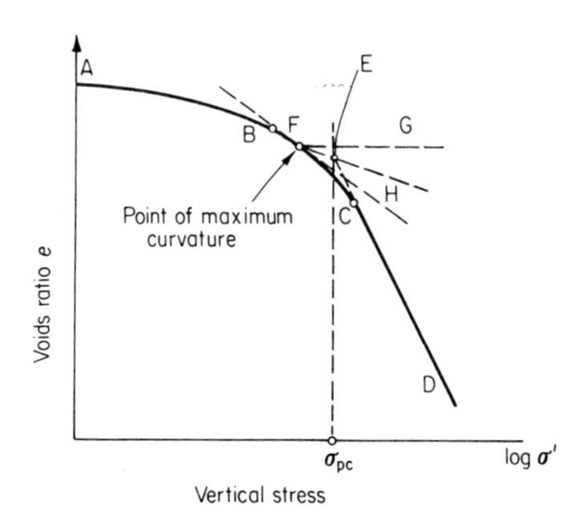

Figure 5.10. Determination of preconsolidation stress.

In the case of reloading of preconsolidated soils (overconsolidated soils) the compression curve in a semilogarithmic scale is as shown in Figure 5.10. The part AB of the curve shows the reloading range while the straight part CD describes a virgin loading range; this implies that the soil has, in its geological history, been consolidated to a point somewhere between B and C.

Casagrande (1936) has suggested an empirical graphical method for determining the preconsolidation stress: firstly point F is found corresponding to the maximum curvature of the section BC; then through the point F a tangent FH and a line FG parallel to the stress axis are drawn; the point of intersection E of the bisector of the angle GFH and the continuation of the straight part of the curve CD gives the value of the maximum consolidation stress in the soil's geological history.

It must be emphasized that the above method only gives an estimate of the maximum preconsolidation stress σ_{pc} which could have resulted either from the weight of overburden or from desiccation, i.e. drying out of soils.

5.2.5. DEFORMATION MODULUS AND COEFFICIENT OF LATERAL STRESS

In unconfined axial compression of soil sample, its axial deformation will always be greater than in the case of the confined (oedometric) compression (Figure 5.11).

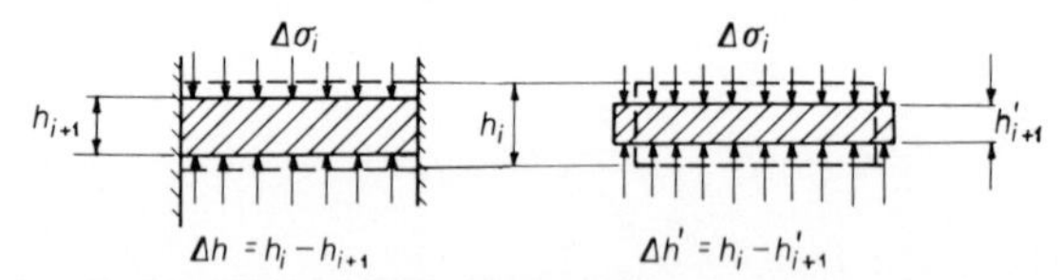

Figure 5.11. Deformation of confined and unconfined soil samples in one-dimensional compression.

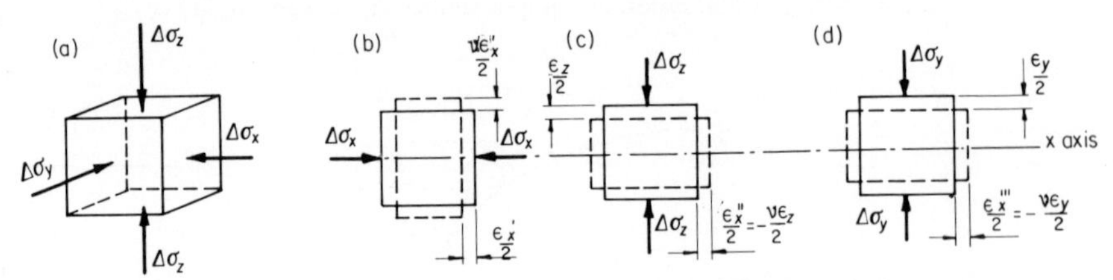

Figure 5.12. Deformation of soil element in the direction of x axis due to stress increments $\Delta\sigma_x$, $\Delta\sigma_y$, and $\Delta\sigma_z$.

Assuming, as before, that for small stress increments the stress–strain relationship is linear, a deformation modulus for laterally unrestrained compression can be obtained:

$$E_v = \Delta\sigma_i h_i / \Delta h'_i$$

The value of the deformation modulus E_v is always smaller than that of the stiffness modulus E' because $\Delta h'_i$ is always greater than Δh_i.

The moduli E_v and E' are interrelated and for the case of isotropic soils this relationship can be obtained as follows. During compression of a soil sample in an oedometer the lateral strains are suppressed by the confining ring in which the sample is placed. As the result of this horizontal stresses $\Delta\sigma_x$ and $\Delta\sigma_y$ are induced in the sample (Figure 5.12). These stresses are equal and are proportioned to the vertical stress increment $\Delta\sigma_z$:

$$\Delta\sigma_x = \Delta\sigma_y = K\Delta\sigma_z \tag{5.6}$$

The constant K above is known as the incremental coefficient of lateral stress. To determine its value, let us consider the strains of the cubical element in the x direction due to the individual action of stress increments $\Delta\sigma_x$, $\Delta\sigma_z$, and $\Delta\sigma_y$ (Figure 5.12). It shall be assumed that deformations in the direction of action of $\Delta\sigma_x$ are positive and in the opposite direction are negative.

The deformations are evaluated using the deformation modulus E_ν, because it is assumed that at any time the cubical element is acted upon by one stress component only and is free to deform laterally to its line of action.

Owing to the action of stress increment $\Delta\sigma_x$ the compressive strain in the x direction (Figure 5.12(b)) is

$$\epsilon_x' = \Delta\sigma_x/E_\nu \tag{a}$$

while, owing to the action of stress increments $\Delta\sigma_z$ and $\Delta\sigma_y$, the strains in the same direction (Figure 5.12(c), (d)) are

$$\epsilon_x'' = -\nu\Delta\sigma_z/E_\nu \tag{b}$$

and

$$\epsilon_x''' = -\nu\Delta\sigma_y/E_\nu \tag{c}$$

where ν = Possion's

For zero lateral strain

$$\epsilon_x' + \epsilon_x'' + \epsilon_x''' = 0$$

or

$$(\Delta\sigma_x - \nu\Delta\sigma_y - \nu\Delta\sigma_z)/E_\nu = 0 \tag{d}$$

Substituting $\Delta\sigma_x = \Delta\sigma_y = K\Delta\sigma_z$ into (d)

$$(K - \nu K - \nu)\Delta\sigma_z/E_\nu = 0$$

and hence

$$K = \nu/(1 - \nu) \tag{5.7}$$

Let us now consider the vertical deformation of a soil sample (in an oedometer) of initial thickness h_1 and under the action of the three stress increments $\Delta\sigma_x$, $\Delta\sigma_y$, and $\Delta\sigma_z$:

$$\Delta h = (\Delta\sigma_z - \nu\Delta\sigma_x - \nu\Delta\sigma_y)h_1/E_\nu = h_1\Delta\sigma_z(1 - 2K\nu)/E_\nu$$

Substituting now for K from Equation (5.7)

$$\Delta h = h_1 \Delta \sigma_z (1+\nu)(1-2\nu)/E_\nu(1-\nu)$$

or

$$\Delta h = \delta h_1 \Delta \sigma_z / E_\nu \qquad \text{(e)}$$

where

$$\delta = (1+\nu)(1-2\nu)/(1-\nu) \qquad (5.8)$$

Vertical deformation of the same sample in terms of the stiffness modulus E' can be determined from Equation (5.1):

$$\Delta h = h_1 \Delta \sigma_z / E' \qquad \text{(f)}$$

Equating the right-hand sides of the equations (e) and (f),

$$E_\nu = \delta E' \qquad (5.9)$$

Values of ν, K, and δ can be taken from Table 5.3, in which values of ν and K have been taken according to Litvinov (1951).

Table 5.3. Average values of coefficients ν, K, and δ for different soils

Type and consistency of soil	ν	K	δ
dense sand	0·25	0·33	0·84
loose sand	0·30	0·43	0·74
clayey sands and silts	0·30	0·43	0.74
firm and stiff sand – or silt–clays	0·35	0·54	0·63
firm and stiff clays	0·40	0·67	0·47
very stiff clays	0·20	0·25	0·90

The above values apply to static loading; if dynamic loading is anticipated then for loose sands and made up ground the authors suggest that the values of K should be increased by about 50 to 70%.

5.2.6. COMPRESSIBILITY OF METASTABLE SOILS

Owing to deposition processes (Chapter 1) some cohesive soils such as loess may have a very open, weakly cemented macroporous skeletal structure and hence may, on being saturated, exhibit considerable additional settlements.

The compressibility of metastable soils is determined from routine oedometer tests in which the initial stages of loading are carried out without filling the cell with water. On completion of consolidation under a stress of 300 kN/m^2 (3·0 tonf/ft^2, 3·0 kgf/cm^2) the cell is filled by passing the water upwards through the sample and hence saturating the soil ($S_r = 1{\cdot}0$).

The index of macroporosity I_m is then determined:

$$I_m = (h - h')/h \qquad (5.10)$$

where h = final thickness of soil sample consolidated under the pressure of 300 kN/m^2 without water in the cell

h' = final thickness of soil sample under the same pressure but after its complete saturation with water ($S_r = 1{\cdot}0$) (Figure 5.13)

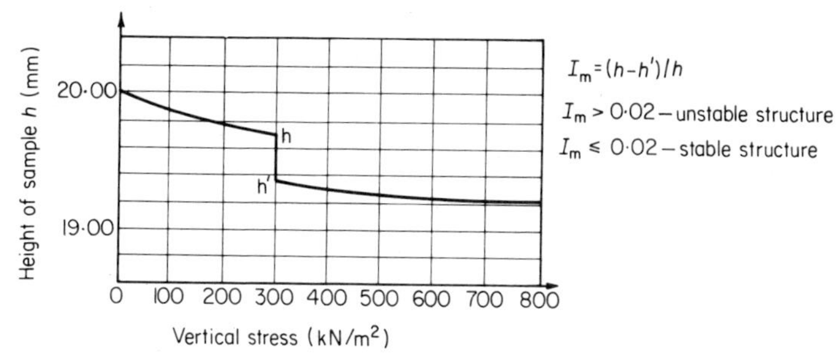

Figure 5.13. Typical compression curve of a metastable soil.

Macroporous soils having the index of macroporosity $I_m \leqslant 0{\cdot}02$ are considered to have a stable structure, while those with $I_m > 0{\cdot}02$ are classed as having an unstable structure.

If a soil is found to have an unstable structure, then it is recommended that another sample of the same soil should be tested but this time with the initial consolidation stress equal to the proposed working stress and not to the standard 300 kN/m^2. Unless great care is taken in preparation of loess samples and in placing them in the confining ring and between the porous stones, then unrealistically high additional settlements are going to be observed on saturation of samples which will lead to erroneous predictions of the field settlements.

5.2.7. DEFORMATION MODULI IN TRIAXIAL COMPRESSION

The use of the oedometer test for determination of the compressibility of soils is well established and there is a considerable amount of field evidence available to confirm that for the majority of routine foundation problems the predicted settlements are well within acceptable limits of accuracy. There are, however, several limitations inherent in this apparatus which may limit its use in detailed studies of stress deformation characteristics required for the more comprehensive settlement analyses of special foundation problems. The most important of these limitations are the following: (1) under zero lateral strain condition there is no control over the lateral stress and hence only one possible combination of axial and lateral stresses can be applied to the tested soil; (2) it is not possible to investigate undrained stress–strain behaviour; the suppression of all shear strains (except those accompanying volumetric strain) ensures that the pore pressure always increases by an amount equal to the vertical stresses until drainage has commenced and hence maintains a constant level of effective stress; (3) there is no control over drainage and no facility for measurement of the pore water pressure; (4) large measurement errors in connection with the small thickness of samples.

The above limitations have led to the adaptation of the more versatile triaxial compression apparatus for the study of the stress deformation characteristics (Skempton and Bjerrum, 1957; Skempton and Sowa, 1963; Lambé, 1967; Davis and Poulos, 1968; Kérisel and Quatre, 1968; Simons and Som, 1969; and many others) and to the development of a hydraulic oedometer (Rowe and Barden, 1966) in which the drainage and its direction can be controlled and pore water pressure measurements taken.

It must be emphasized at this stage that the apparatus and the expertise required for most of the tests described by the above authors are at present only available in small number of university, government, and industrial research laboratories and, for a very long time to come, will remain there. The principles behind this general approach are, however, basic, and so well illustrate the present trends in settlement analysis, that they are worth reviewing.

Let us consider a small cylindrical element of soil vertically below a point A on the surface of a soil half-space (Figure 5.14(a)). The *in situ* effective stresses acting on the soil, prior to application of any loading, are the vertical overburden stress σ'_{oz} and the horizontal radial stress σ'_{or} which are interdependent and can normally be taken to be linearly proportional, i.e.

$$\sigma'_{or} = K_0 \sigma'_{oz}$$

where K_0 is the constant of proportionality known as the coefficient of lateral stress at rest which for normally consolidated soils varies between 0·4 and 0·7 and for overconsolidated soils can be as high as 2·5 (Skempton, 1960).

The pore water pressure within the element is

$$u_o = (g\rho_w)h$$

or

$$u_o = \gamma_w h$$

Therefore the total *in situ* stresses on the element are

$$\sigma_{oz} = \sigma'_{oz} + u_o$$

and

$$\sigma_{or} = K_0 \sigma'_{oz} + u_o$$

Let us now assume that application of surface loading at A induces only axial and radial stress increments on the element (Figure 5.14(b)) which according to Boussinesq's theory are equal to $\Delta\sigma_z$ and $\Delta\sigma_r$* and which are accompanied by axial and radial strains ϵ_z and ϵ_r. In order to be able to predict

* For the sake of clarity the effects of the changes in Poisson's ratio during the consolidation process on the radial stresses have been omitted from the foregoing discussion.

these strains it is necessary to know the compressibility of the soil under the particular combination of the stress increments $\Delta\sigma_z$ and $\Delta\sigma_r$ and not under the zero lateral strain condition as in the oedometer test. Further more, because the lateral (radial) strains are not suppressed, the pore water pressure induced in the soil element by the loading, prior to the commencement of drainage, will not be equal to the vertical stress increment $\Delta\sigma_z$ but will be given by the following equation (Skempton and Sowa, 1963):

$$\Delta u = B\left\{\frac{\Delta\sigma_z + 2\Delta\sigma_r}{3} + (A - \tfrac{1}{3})|\Delta\sigma_z - \Delta\sigma_r|\right\} \qquad (5.11)$$

where Δu = change in pore water pressure in undrained condition

B = pore pressure coefficient dependent on the degree of saturation of soil (S_r) and equal to 1·0 for S_r = 1·0

A = pore pressure coefficient which for any given soil depends on the magnitude of the deviatoric stress ($\Delta\sigma_z - \Delta\sigma_r$) and for small values of deviatoric stress is approximately equal to $\frac{1}{3}$

$|\Delta\sigma_z - \Delta\sigma_r|$ = absolute value of deviatoric stress

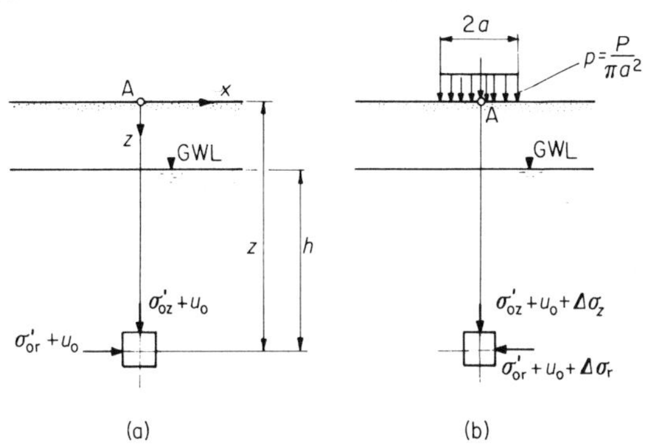

Figure 5.14. Vertical and radial stresses on cylindrical soil element at depth z: (a) before application of loading; (b) after application of surface loading.

For small deviatoric stresses the induced pore water pressure Δu obtained from Equation (5.11) will approximately be equal to the mean applied stress and hence will be smaller than the axial stress increment $\Delta\sigma_z$ but greater than the radial stress increment $\Delta\sigma_r$. This will result in an immediate increase in the effective axial stress $\Delta\sigma_z - \Delta u$ and a decrease in the effective radial stress $\Delta\sigma_r - \Delta u$ which must be accompanied by axial compression and radial extension of the element; the former leads to the *immediate settlement* of the foundation which will occur without any volumetric strain of the soil, i.e. prior to the commencement of the consolidation process.

During the subsequent dissipation of the excess pore water pressure ($\Delta u \to 0$) the increase in the effective stresses acting on the element is accompanied by a decrease in its volume and hence by further axial and radial compressive strains which lead to the *consolidation settlement* of the foundation; the rate of this settlement is governed by the consolidation process, i.e. it depends on the porosity and permeability of the soil and on the nature of the interparticle bonds.

The above-described process of *in situ* loading of a soil element can be reproduced in a triaxial compression test (Section 5.4.5) and from its results both the undrained and drained stress deformation characteristics of the soil can be determined.

Before the effects of the induced stresses ($\Delta\sigma_z$ and $\Delta\sigma_r$) can be investigated it is essential to ensure that the porosity, water content, and the initial stresses in the soil sample resemble, as closely as possible, the *in situ* conditions (Figure 5.14(a) and Figure 5.15, Stage 1). Assuming that mechanical disturbance

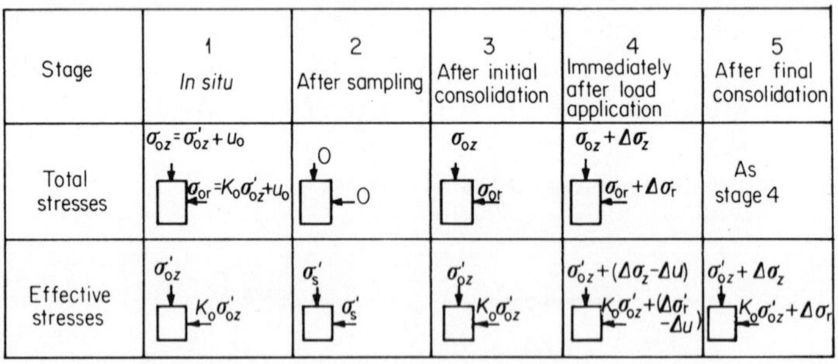

Stage	1 *In situ*	2 After sampling	3 After initial consolidation	4 Immediately after load application	5 After final consolidation
Total stresses	$\sigma_{oz} = \sigma'_{oz} + u_0$; $\sigma_{or} = K_0\sigma'_{oz} + u_0$	0; 0	σ_{oz}; σ_{or}	$\sigma_{oz} + \Delta\sigma_z$; $\sigma_{or} + \Delta\sigma_r$	As stage 4
Effective stresses	σ'_{oz}; $K_0\sigma'_{oz}$	σ'_s; σ'_s	σ'_{oz}; $K_0\sigma'_{oz}$	$\sigma'_{oz} + (\Delta\sigma_z - \Delta u)$; $K_0\sigma'_{oz} + (\Delta\sigma'_r - \Delta u)$	$\sigma'_{oz} + \Delta\sigma_z$; $K_0\sigma'_{oz} + \Delta\sigma_r$

Figure 5.15. Total and effective stresses on a soil sample at various stages prior to and during testing.

of the soil structure and any change in the water content of the soil can be avoided in the process of extraction and preparation of the sample, then, the only 'sampling disturbance' requiring consideration will be the change in the state of stress.

On preparation of the sample for testing the external stresses on it are reduced to zero, i.e. the total *in situ* stresses are completely removed $\Delta\sigma_z = -\sigma_{oz}$, $\Delta\sigma_r = -\sigma_{or}$ (Figure 5.15, Stage 2). The removal of these stresses induces a change in the pore water pressure in the sample which can be evaluated from Equation (5.11) and which for small stress changes can be taken as approximately equal to the change in the external effective mean stress:

$$\Delta u_s = -\frac{\sigma'_{oz} + 2\sigma'_{or}}{3} = -\sigma'_{oz}\frac{(1 + 2K_0)}{3}$$

Since there are no external stresses acting on the sample, this change in suction induces, through the surface menisci, an isotropic compression of the soil skeleton:

$$\sigma'_s = -u_s = \sigma'_{oz}\frac{(1 + 2K_0)}{3}$$

In the case of mechanically undisturbed soil sample the magnitude of σ_s' should be equal to the stress required to prevent swelling of the soil when it is in contact with free water (Section 5.2.1) (Skempton, 1960). Hence from the knowledge of σ_s' and σ_{oz}' the coefficient of lateral stress at rest can approximately be evaluated.

With a knowledge of the *in situ* effective stresses and the *in situ* pore water pressure u_o it is now possible to reconsolidate the sample in a triaxial compression apparatus to its *in situ* stress conditions (Figure 5.15, Stage 3); note that for $K_0 > 1{\cdot}0$ the axial stress is less than the radial (cell pressure) and hence it will be necessary to use a special triaxial cell for this type of loading.

Having reinstated the soil to its *in situ* condition its compressibility under the induced stress increments ($\Delta\sigma_z$ and $\Delta\sigma_r$) will now be investigated in two phases: (1) in undrained condition to determine the immediate (undrained) compressibility of the soil, and (2) with free drainage to determine consolidation deformations which when combined with those from the first phase will enable the determination of the overall compressibility of the soil.

In the undrained phase the deformations of the soil sample (Figure 5.16(a)) occur without any change in volume; the isotropic undrained deformation

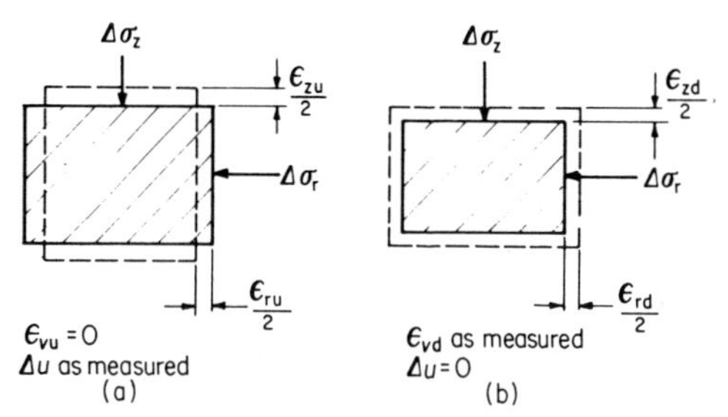

Figure 4.16. Deformation of soil element due to application of stress increments, (a) undrained phase (b) final drained phase.

modulus E_u can now be evaluated in terms of the known stress increments (Figure 5.15, Stage 4) and the measured strains:

$$E_u \epsilon_{zu} = \Delta\sigma_z - 2\nu_u \Delta\sigma_r$$

but, for no volume change, Poisson's ratio ν_u is equal to $0{\cdot}5$ and hence

$$E_u = (\Delta\sigma_z - \Delta\sigma_r)/\epsilon_{zu} \tag{5.12}$$

In order to minimize creep effects under the deviatoric undrained loading the second phase of the test is started as soon as the induced pore water pressure has become constant.

As pore water is allowed to drain at the start of the second phase of the test changes in volume and height of the sample are recorded at the same time intervals as in the oedometer test (Section 5.2.1); the former by measuring the volume of water expelled from the sample and the latter in the usual manner. Consolidation curves plotted in a semilogarithmic scale are similar to that shown in Figure 5.5 and the end of the primary consolidation is taken as the end of the test (Figure 5.16(b)). By the end of the consolidation stage the effective stresses in the soil are equal to the applied stresses (Figure 5.15, Stage 5).

The results of the two phases are now combined to give the total axial, radial, and volumetric strains (ϵ_{zt}, ϵ_{rt}, and ϵ_{vt}) which together with the applied stress increments are now used to determine the overall isotropic deformation modulus E_ν and the corresponding Poisson's ratio ν.

From Figures 5.16(a) and 5.16(b) the total axial strain ϵ_{zt} is

$$\epsilon_{zt} = \epsilon_{zu} + \epsilon_{zd}$$

while the total volumetric strain is

$$\epsilon_{vt} = 0 + \epsilon_{vd}$$

Combining the above results the total radial strain ϵ_{rt} is obtained:

$$\epsilon_{rt} = (\epsilon_{vt} - \epsilon_{zt})/2$$

Expressing now the axial and radial strains in terms of E_ν and ν and the applied stress increments $\Delta\sigma_z$ and $\Delta\sigma_r$:

$$E_\nu \epsilon_{zt} = \Delta\sigma_z - 2\nu\Delta\sigma_r$$
$$E_\nu \epsilon_{rt} = \Delta\sigma_r (1 - \nu) - \nu\Delta\sigma_z$$

and solving the two simultaneous equation the isotropic overall deformation modulus E_ν and Poisson's ratio ν are obtained in terms of the known stresses and strains:

$$E_\nu = (\Delta\sigma_z - 2\nu_t\Delta\sigma_r)/\epsilon_{zt} \tag{5.13}$$

and

$$\nu = \frac{\epsilon_{zt}\Delta\sigma_r - \epsilon_{rt}\Delta\sigma_z}{\epsilon_{zt}(\Delta\sigma_z + \Delta\sigma_r) - 2\epsilon_{rt}\Delta\sigma_r} \tag{5.14}$$

Both the above deformation moduli are of course secant moduli which are only applicable to the particular combination of the applied stress increments $\Delta\sigma_z$ and $\Delta\sigma_r$.

5.3. Rate of Deformation of Soils

The preceding sections have dealt with the total final deformations of soils which are reached on the completion of the time-dependent process.

This process which, as we saw, involves a decrease in the volume of pores or water content due to an increase in applied stresses is referred to as *consolidation of soils*; the increase in the volume of soils due to a decrease in applied stresses is referred to as *expansion*.

The rate of consolidation depends on the permeability (in the case of full saturation), on the creep characteristics of the soil skeleton, and also on the deformation characteristics of the other numerous constituent materials which may be present in natural soils (pore water, entrapped air, water vapour and gases, organic matter, etc.).

The soils which present foundation engineers with the most difficult settlement problems are usually the recent, fully saturated, alluvial deposits. Settlements of structures or embankments founded on such soils can be of the order of several metres, and may take hundreds of years to complete (e.g. the Tower of Piza). In further consideration we shall, therefore, limit ourselves to fully saturated soils which may or may not be of alluvial origin.

The process of deformation of saturated soils on application of external loading can be divided into three phases: the initial *instantaneous deformation* (adiabatic, governed by the compressibility of the pore water), the process of *primary consolidation* associated with the dissipation of excess pore water pressures with the consequent expulsion (drainage) of water from the soil, and, finally, the process of *secondary consolidation* or *creep deformation* of the soil skeleton, associated with the deformation of the adsorbed water layers and slippage between particles, during which changes in water content are negligible.

At present, the most widely accepted theory of consolidation, which is used in determination of the rate of deformation of fully saturated soils, is limited to the *primary consolidation (drainage) process.*

5.3.1. ASSUMPTIONS IN THE THEORY OF CONSOLIDATION

(1) The pores in the soil are assumed to be hydraulically continuous and completely filled with free, incompressible, water.

(2) The soil skeleton is assumed to deform linearly and the deformation is instantaneous on the application of effective stress.

(3) The soil is homogeneous and the external loading applied to it is, in the first instance, wholly carried by the pore water.

(4) The flow (drainage) of the water in the pores obeys Darcy's law.

The above assumptions form the basis of the theory of consolidation developed in the following section which, without modifications, is generally only applicable to normally consolidated, fully saturated, soft cohesive soils.

5.3.2. ONE-DIMENSIONAL THEORY OF CONSOLIDATION

Uniform consolidation of a layer of soil of finite thickness, extending indefinitely in plan and, for example, sandwiched between two layers of permeable soils (sand or gravel) is known as one-dimensional consolidation. A sample of soil consolidated in an oedometer between two porous stones can be considered as a model of such a layer. In order to develop the theory of one-dimensional consolidation let us consider the process of consolidation of a layer shown in

Figure 5.17 in which the distribution of the total stress due to the applied loading p_1 is uniform.

On an instantaneous application of the loading to a layer of saturated soil, the entire load is initially transmitted to the pore water. However, the very steep hydraulic gradient set up within the pore water at the free boundary induces an immediate flow of water towards the boundary and leads to a

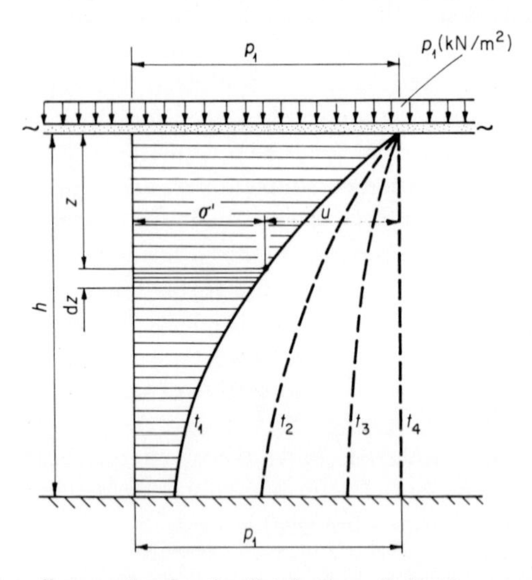

Figure 5.17. Stress diagram showing the distribution of effective stresses σ' and pore water pressures u at various times, in a saturated layer of soil under uniform load increment p_1.

gradual transfer of the load from the pore water to the soil skeleton; at any instant and at any point within the soil the sum of the effective stress in the soil σ' and the pore water pressure u is equal to the applied stress p_1, i.e.

$$p_1 = \sigma' + u \tag{a}$$

Since the soil under consideration is saturated, the increase in the flow of water from an elemental layer, in any time interval, must be equal to the decrease in the porosity of the soil, i.e. it can be assumed that

$$\frac{\partial v_z}{\partial z} = -\frac{\partial n}{\partial t} \tag{b}$$

where v_z = velocity of flow of water
n = porosity, i.e. the volume of voids per unit volume of soil
t = time

The above differential equation which governs the one-dimensional consolidation of soils was first derived by Terzaghi in 1925.

Let us now separately consider both sides of the equation (b). Applying Darcy's law to the left-hand side of the equation we can write

$$v_z = -k \frac{\partial H}{\partial z} \tag{c}$$

and, on differentiation,

$$\frac{\partial v_z}{\partial z} = -k \frac{\partial^2 H}{\partial z^2} \tag{d}$$

Assuming that the excess hydrostatic head H is equal to the excess pore water pressure u, divided by the unit weight of water γ_w, and utilizing the equation (a) we obtain

$$u = p_1 - \sigma'; \; H = \frac{u}{\gamma_w} \text{ or } H = \frac{p_1 - \sigma'}{\gamma_w}$$

and therefore

$$\frac{\partial^2 H}{\partial z^2} = -\frac{1}{\gamma_w} \times \frac{\partial^2 \sigma'}{\partial z^2} \tag{e}$$

On substitution into (d) we obtain

$$\frac{\partial v_z}{\partial z} = \frac{k}{\gamma_w} \times \frac{\partial^2 \sigma'}{\partial z^2} \tag{f}$$

By assuming on the right-hand side of (b) that porosity $n = e/(1 + e)$ and neglecting in the denominator the change in the voids ratio (which is small in comparison with 1·0), by replacing it with its initial value e_1, we obtain

$$\frac{\partial n}{\partial t} = \frac{1}{1 + e_1} \times \frac{\partial e}{\partial t} \tag{g}$$

Utilizing Equation (5.2) (with a change of sign to allow for the negative slope of the (e, σ') curve) we can write that

$$\frac{\partial e}{\partial t} = -a_v \frac{\partial \sigma'}{\partial t} \tag{h}$$

and, therefore, the right-hand side of (b) can be written in terms of the rate of change of the effective stress:

$$\frac{\partial n}{\partial t} = -\frac{a_v}{1 + e_1} \times \frac{\partial \sigma'}{\partial t} \tag{i}$$

According to Equation (5.4) the constant term $a_v/(1 + e_1) = m_v$, the coefficient of volume compressibility and the equation (i) simplifies to

$$\frac{\partial n}{\partial t} = -m_v \frac{\partial \sigma'}{\partial t} \tag{j}$$

On substitution now of the above expressions for $\partial v_z/\partial z$ and $\partial n/\partial t$ in (b) and on grouping all the constant parameters on the left-hand side, we obtain

$$\frac{k}{m_v \gamma_w} \cdot \frac{\partial^2 \sigma'}{\partial z^2} = \frac{\partial \sigma'}{\partial t} \tag{k}$$

By replacing the term containing all the constant parameters with the *coefficient of consolidation* c_v,

$$c_v = \frac{k}{m_v \gamma_w} \tag{5.15}$$

we finally obtain

$$c_v \frac{\partial^2 \sigma'}{\partial z^2} = \frac{\partial \sigma'}{\partial t} \tag{5.16}$$

Note in Equation (5.15) that, although k and m_v are stress dependent, experimental evidence indicates that their ratio is fairly constant.

This is the governing differential equation of the one-dimensional consolidation process.

By referring back to (a) we can obtain the governing equation in terms of the excess pore water pressure:

$$c_v \frac{\partial^2 u}{\partial z^2} = \frac{\partial u}{\partial t} \tag{5.17}$$

The solution of a partial differential equation depends on the boundary conditions. For the case under consideration or for the case of a layer of soil of thickness $2h$ but with free drainage at both boundaries (top and bottom) the solution is given by the Fourier series

$$u = \frac{4}{\pi} p_1 \left\{ \sin\left(\frac{\pi z}{2h}\right) \times e^{-N} + \frac{1}{3} \sin\left(\frac{3\pi z}{2h}\right) \times e^{-9N} + \frac{1}{5} \sin\left(\frac{5\pi z}{2h}\right) \times e^{-25N} + \ldots \right\} \tag{5.18}$$

where

$$N = \frac{\pi^2 c_v}{4h^2} t \tag{5.19}$$

Note that in many references a dimensionless parameter known as the time factor T_v is used

$$T_v = \frac{c_v}{h^2} t = \frac{4}{\pi^2} N \tag{5.20}$$

If knowledge of the effective stress is required, as may be the case if, for example, the variation of the shear strength τ_f at depth $z = h$ is being investigated, then, utilizing Equation (5.18) and neglecting higher terms we can write that

$$\sigma'_h = p_1 - u \approx p_1 \left(1 - \frac{4}{\pi} e^{-N}\right) \tag{5.21}$$

and on substitution in the expression for the shear strength we obtain

$$\tau_f = c' + \sigma'_h \tan \phi' = c' + p_1 \left(1 - \frac{4}{\pi} e^{-N}\right) \tan \phi' \tag{5.22}$$

In practice, however, we are mainly concerned with the progress of consolidation of a given layer of soil after the application of loading, i.e. with the settlement at any time t.

The progress of consolidation at any time may be determined by comparing the settlement at time t, s_t, with the final consolidation settlement, s_f. The *degree of consolidation* U_c is defined by

$$U_c = \frac{s_t}{s_f} \tag{5.23}$$

Because the consolidation settlement is directly proportional to the effective stresses we can also define the degree of consolidation as the ratio of the area of the effective stress diagram at any time, t, to the area of the final effective stress diagram (at $t = \infty$). This can be written in mathematical terms as

$$U_c = \int_0^h \frac{\sigma' \, dz}{A_p} = \int_0^h \frac{(1 - u) \, dz}{A_p} \tag{5.24}$$

where A_p is the area of the final (i.e. at $t = \infty$) effective stress diagram, which for the case under consideration is equal to $p_1 \times h$.

On substitution from Equation (5.18) and integration between the limits, on simplification we obtain

$$U_0 = 1 - \frac{8}{\pi^2} \left(e^{-N} + \frac{1}{9} e^{-9N} + \frac{1}{25} e^{-25N} + \ldots\right) \tag{5.25}$$

where U_0 = degree of consolidation corresponding to the uniform distribution of stress within the consolidating layer of soil

For practical purposes (e.g. for $U_0 < 0{\cdot}4$) only the first term of the series need be considered and then

$$U_0 = 1 - \frac{8}{\pi^2} e^{-N} \tag{5.26}$$

For the problem under consideration we can now write that

$$s_t = U_0 s_f \tag{5.27}$$

and on substitution for s_f from Equations (5.1) and (5.4) and for U_0 from Equation (5.25) we obtain the final expression for the settlement at any time t:

$$s_t = m_v h p_1 \left\{1 - \frac{8}{\pi^2}\left(e^{-N} + \frac{1}{9}e^{-9N} + \frac{1}{25}e^{-25N} + \ldots\right)\right\} \tag{5.28}$$

Example. Determine settlements of a uniformly stressed layer of soil for $t = 1, 2,$ and 5 years, given that

$$h = 5{\cdot}0 \text{ m},\ p_1 = 200 \text{ kN/m}^2,\ m_v = 0{\cdot}1 \text{ m}^2/\text{MN},\ k = 1{\cdot}0 \times 10^{-7} \text{ mm/s}$$

$$\gamma_w = 9{\cdot}8 \text{ kN/m}^3$$

Before we can determine N we must evaluate c_v and for this purpose it is convenient to change units of k and γ_w: 1 mm/s $\approx 3 \times 10^7$ mm/year $= 3 \times 10^4$ m/year and $\gamma_w = 9{\cdot}8$ kN/m$^3 \approx 10^{-2}$ MN/m^3. Therefore

$$c_v = \frac{1{\cdot}0 \times 10^{-7} \times 3 \times 10^4}{0{\cdot}1 \times 10^{-2}} = 3 \text{ m}^2/\text{year}$$

and

$$N = \frac{\pi^2 c_v}{4h^2} t = \frac{9{\cdot}87 \times 3t}{4 \times 25} \approx 0{\cdot}3t$$

The final settlement of the layer is obtained from the following expression:

$$s_f = m_v h p_1 = 0{\cdot}1 \times 5{\cdot}0 \times 200 \times 10^{-3} = 0{\cdot}10 \text{ m} = 100 \text{ mm}$$

For the settlement at $t = 1$ year (using Table 5.4) we obtain

$$e^{-N} = e^{-0{\cdot}3 \times 1} = 0{\cdot}741;\ e^{-9N} = e^{-9 \times 0{\cdot}3 \times 1} = 0{\cdot}067$$

On substitution in Equation (5.28) the settlement at the end of the one year period is

$$s_1 = m_v h p_1 \left\{1 - \frac{8}{\pi^2}\left(e^{-N} + \frac{1}{9}e^{-9N}\right)\right\} = 100\left\{1 - 0{\cdot}81\,(0{\cdot}741 + 0{\cdot}067)\right\} = 39 \text{ mm}$$

Similarly for the settlement at $t = 2$ years: $e^{-N} = e^{-0{\cdot}3 \times 2} = 0{\cdot}549$ and e^{-9N} is small, and therefore

$$s_2 = s_f\left(1 - \frac{8}{\pi^2}e^{-N}\right) = 100\,(1 - 0{\cdot}81 \times 0{\cdot}549) = 56 \text{ mm}$$

For $t = 5$ years

$$s_5 = 100(1 - 0{\cdot}81\, e^{-0{\cdot}3 \times 5}) = 82 \text{ mm}$$

Table 5.4. Values of e^{-x}

x	e^{-x}	x	e^{-x}	x	e^{-x}	x	e^{-x}	x	e^{-x}	x	e^{-x}	x	e^{-x}	x	e^{-x}
0·000	1·0	0·20	0·819	0·50	0·607	0·80	0·449	1·10	0·333	1·40	0·247	1·70	0·183	2·00	0·135
0·001	0·999	0·21	0·811	0·51	0·601	0·81	0·445	1·11	0·330	1·41	0·244	1·71	0·181	2·01	0·134
0·002	0·998	0·22	0·803	0·52	0·595	0·82	0·440	1·12	0·326	1·42	0·242	1·72	0·179	2·02	0·133
0·003	0·997	0·23	0·795	0·53	0·589	0·83	0·436	1·13	0·323	1·43	0·239	1·73	0·177	2·03	0·131
0·004	0·996	0·24	0·787	0·54	0·583	0·84	0·431	1·14	0·320	1·44	0·237	1·74	0·176	2·04	0·130
0·005	0·995	0·25	0·779	0·55	0·577	0·85	0·427	1·15	0·317	1·45	0·235	1·75	0·174	2·05	0·129
0·006	0·994	0·26	0·771	0·56	0·571	0·86	0·423	1·16	0·313	1·46	0·232	1·76	0·172	2·06	0·127
0·007	0·993	0·27	0·763	0·57	0·566	0·87	0·419	1·17	0·310	1·47	0·230	1·77	0·170	2·07	0·126
0·008	0·992	0·28	0·756	0·58	0·560	0·88	0·415	1·18	0·307	1·48	0·228	1·78	0·169	2·08	0·125
0·009	0·991	0·29	0·748	0·59	0·554	0·89	0·411	1·19	0·304	1·49	0·225	1·79	0·167	2·09	0·124
—	—	0·30	0·741	0·60	0·549	0·90	0·407	1·20	0·301	1·50	0·223	1·80	0·165	2·10	0·122
0·01	0·990	0·31	0·733	0·61	0·543	0·91	0·403	1·21	0·298	1·51	0·221	1·81	0·164	2·15	0·116
0·02	0·980	0·32	0·726	0·62	0·538	0·92	0·399	1·22	0·295	1·52	0·219	1·82	0·162	2·20	0·111
0·03	0·970	0·33	0·719	0·63	0·533	0·93	0·394	1·23	0·292	1·53	0·217	1·83	0·160	2·25	0·105
0·04	0·961	0·34	0·712	0·64	0·527	0·94	0·391	1·24	0·289	1·54	0·214	1·84	0·159	2·30	0·100
0·05	0·951	0·35	0·705	0·65	0·522	0·95	0·387	1·25	0·286	1·55	0·212	1·85	0·157	2·35	0·095
0·06	0·942	0·36	0·698	0·66	0·517	0·96	0·383	1·26	0·284	1·56	0·210	1·86	0·156	2·40	0·091
0·07	0·932	0·37	0·691	0·67	0·512	0·97	0·379	1·27	0·281	1·57	0·208	1·87	0·154	2·45	0·086
0·08	0·923	0·38	0·684	0·68	0·507	0·98	0·375	1·28	0·278	1·58	0·206	1·88	0·152	2·50	0·082
0·09	0·914	0·39	0·677	0·69	0·502	0·99	0·372	1·29	0·275	1·59	0·204	1·89	0·151	2·55	0·078
0·10	0·905	0·40	0·670	0·70	0·497	1·00	0·368	1·30	0·273	1·60	0·202	1·90	0·150	2·6	0·074
0·11	0·896	0·41	0·664	0·71	0·492	1·01	0·364	1·31	0·270	1·61	0·200	1·91	0·148	2·7	0·067
0·12	0·887	0·42	0·657	0·72	0·487	1·02	0·351	1·32	0·267	1·62	0·198	1·92	0·147	2·8	0·061
0·13	0·878	0·43	0·651	0·73	0·482	1·03	0·357	1·33	0·264	1·63	0·196	1·93	0·145	2·9	0·055
0·14	0·869	0·44	0·644	0·74	0·477	1·04	0·353	1·34	0·262	1·64	0·194	1·94	0·144	3·0	0·050
0·15	0·861	0·45	0·638	0·75	0·472	1·05	0·350	1·35	0·259	1·65	0·192	1·95	0·142	4·0	0·018
0·16	0·852	0·46	0·631	0·76	0·467	1·06	0·346	1·36	0·257	1·66	0·190	1·96	0·141	5·0	0·007
0·17	0·844	0·47	0·625	0.77	0·463	1·07	0·343	1·37	0·254	1·67	0·188	1·97	0·140	6·0	0·002
0·18	0·835	0·48	0·619	0·78	0·458	1·08	0·340	1·38	0·252	1·68	0·186	1·98	0·138	7·0	0·001
0·19	0·827	0·49	0·613	0·79	0·454	1·09	0·336	1·39	0·249	1·69	0·185	1·99	0·137	10·0	0·000

The time–settlement curve for this example is shown in Figure 5.18.

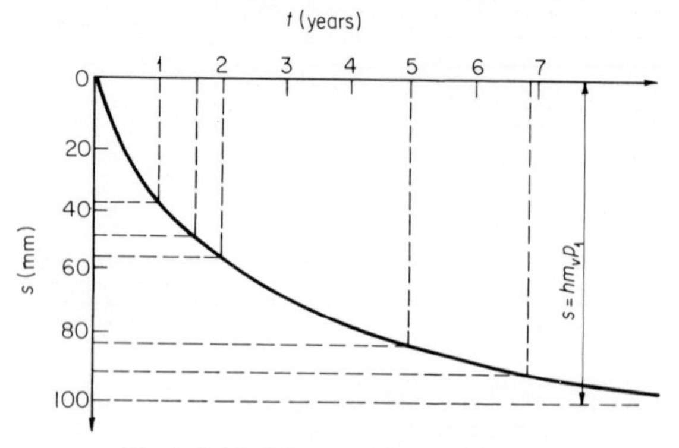

Figure 5.18. Time–settlement curve.

So far we have considered the problem of one-dimensional consolidation of a layer of soil in which the distribution of the total stress due to the applied loading was uniform, i.e. the total stress diagram was rectangular (Figure 5.19(a)). This particular problem will be referred to as *case 0.* Other one-dimensional consolidation problems which are of practical interest are *case 1*, in which the

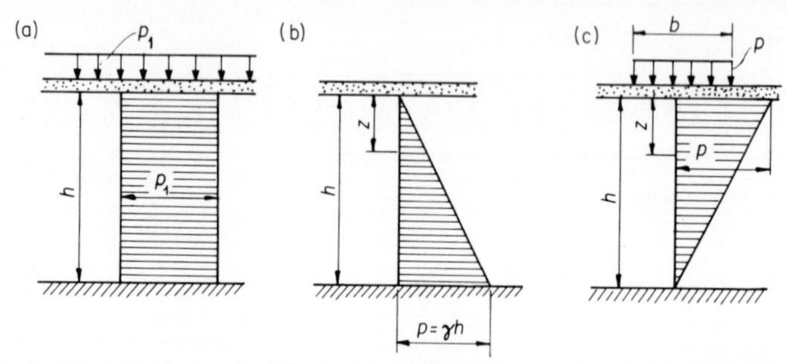

Figure 5.19. Different cases of distribution of stresses in the one-dimensional consolidation problem: (a) case 0; (b) case 1; (c) case 2.

distribution of the total stress due to the applied loading is triangular and increases with depth (Figure 5.19(b)); *case 2*, in which the distribution of the total stress due to the applied loading is triangular and decreases with depth (Figure 5.19(c)) and the *combined case* in which the distribution of the total stress due to the applied loading is trapezoidal and either increases or decreases with depth.

Case 1—a linear increase of stress with depth will occur, for example, in the case of a newly placed soil which is consolidating under its own weight, then

$$\sigma_z = \gamma h \frac{z}{h} = p \frac{z}{h} \quad \text{(Figure 5.19(b))}$$

The solution of Equation (5.17) for the given boundary conditions ($u = 0$ at $z = 0$ and $du/dz = 0$ at $z = h$) will enable us to determine an expression for the excess pore pressure u and hence for the degree of consolidation U_1, which will be equal to

$$U_1 = 1 - \frac{32}{\pi^3}\left(e^{-N} - \frac{1}{27}e^{-9N} + \frac{1}{125}e^{-25N} \pm \ldots\right) \quad (5.29)$$

Therefore, the settlement at any time t of a layer of soil, which consolidates under stresses which increase with depth (assuming that the average total stress is equal to $p/2$) will be given by the following expression:

$$s_{1t} = \frac{m_v hp}{2}\left\{1 + \frac{32}{\pi^3}\left(e^{-N} - \frac{1}{27}e^{-9N} \pm \ldots\right)\right\} \quad (5.30)$$

Case 2 approximates in practice to the stress distribution in a thick layer of soil under a strip foundation and, therefore, is of considerable practical interest; the stress at any depth z is

$$\sigma_z = p\left(1 - \frac{z}{h}\right)$$

On solution of Equations (5.17) and (5.24) we obtain

$$U_2 = 1 - \frac{16}{\pi^2}\left\{\left(1 - \frac{2}{\pi}\right)e^{-N} + \frac{1}{9}\left(1 + \frac{2}{3\pi}\right)e^{-9N} + \ldots\right\} \quad (5.31)$$

and settlement at any time t is given by the expression

$$s_{2t} = \frac{m_v hp}{2}\left[1 - \frac{16}{\pi^2}\left\{\left(1 - \frac{2}{\pi}\right)e^{-N} + \frac{1}{9}\left(1 + \frac{2}{3\pi}\right)e^{-9N} + \ldots\right\}\right] \quad (5.32)$$

By combining the above expressions for the degree of consolidation obtained for the three different cases of stress distribution it can be shown that

$$U_2 = 2U_0 - U_1 \quad (5.33)$$

Equation (5.33) enables one to determine U_2 from a knowledge of U_0 and U_1 without the evaluation of Equation (5.31).

To simplify calculations the values of N corresponding to different degrees of consolidation U_c have been tabulated (Table 5.5) for the three cases of stress distribution: uniform (case 0) and triangular (cases 1 and 2).

If the distribution of stresses in the layer of soil under consideration is approximately trapezoidal, then the values of U_c and N are obtained by interpolation using the tabulated values of N for cases 0 and 1 (when stress increases with depth) and for cases 0 and 2 (when stress decreases with depth).

Table 5.5. Values of N for determination of rate of settlement

$U = \frac{s_t}{s_f}$	Value of N for case			$U = \frac{s_t}{s_f}$	Value of N for case		
	0	1	2		0	1	2
0·05	0·005	0·06	0·002	0·55	0·59	0·84	0·32
0·10	0·02	0·12	0·005	0·60	0·71	0·95	0·42
0·15	0·04	0·18	0·01	0·65	0·84	1·10	0·54
0·20	0·08	0·25	0·02	0·70	1·00	1·24	0·69
0·25	0·12	0·31	0·04	0·75	1·18	1·42	0·88
0·30	0·17	0·39	0·06	0·80	1·40	1·64	1·08
0·35	0·24	0·47	0·09	0·85	1·69	1·93	1·36
0·40	0·31	0·55	0·13	0·90	2·09	2·35	1·77
0·45	0·39	0·63	0·18	0·95	2·80	3·17	2·54
0·50	0·49	0·73	0·24	1·00	∞	∞	∞

The values of the interpolation coefficients I and I' are given in Table 5.6 and depend on α, the ratio of the stresses at $z = 0$ and $z = h$.

The value of N for trapezoidal stress distributions are obtained from the following expressions:

for case 0–1
$$N_{0-1} = N_0 + (N_1 - N_0)I \quad (5.34)$$
and
for cases 0–2
$$N_{0-2} = N_2 + (N_0 - N_2)I' \quad (5.35)$$

Table 5.6. Coefficients I and I'

Case	α	0	0·1	0·2	0·3	0·4	0·5	0·6	0·7	0·8	0·9	1·0
0–1	I	1	0·84	0·69	0·56	0·46	0·36	0·27	0·19	0·12	0·06	0
Case	α	1	1·5	2·0	2·5	3·0	3·5	4	5	7	10	20
0–2	I'	1	0·83	0·71	0·62	0·55	0·50	0·45	0·39	0·30	0·23	0·13

In evaluations of settlement using the above tables, the degree of consolidation is first assumed (e.g. $U_c = 0{\cdot}2, 0{\cdot}4, 0{\cdot}6$, etc.) and, after the value of N has been obtained (either directly from Table 5.5 or by interpolation), the time t corresponding to the assumed degree of consolidation is evaluated from Equation (5.19):

$$t = \frac{4h^2}{\pi^2 c_v} N \quad (5.36)$$

Note that the results obtained using Equations (5.28) to (5.32) are identical with those obtained using the above tables.

Example. Using the data from the previous example, determine the settlements and the corresponding times required to achieve 50% and 90% consolidation (i.e. $U_c = 0{\cdot}5$ and $0{\cdot}9$).

The final settlement worked out in the previous example was $s_f = 100$ mm and the coefficient of consolidation, $c_v = 3\ m^2/year$.

For $U_0 = 0{\cdot}5$ settlement $s_t = 0{\cdot}5 \times 100 = 50$ mm, and from Table 5.5 the corresponding $N_0 = 0{\cdot}49$.

Therefore

$$t_{50} = \frac{4h^2}{\pi^2 c_v} N_0 = \frac{4 \times 25}{9{\cdot}87 \times 3} \times 0{\cdot}49 = 1{\cdot}6 \text{ years}$$

In the same way for the degree of consolidation $U_0 = 0{\cdot}9$, $s_t = 0{\cdot}9 \times 100 = 90$ m and $N_0 = 2{\cdot}09$, and hence the time to achieve it

$$t_{90} = \frac{4 \times 25}{9{\cdot}87 \times 3} \times 2{\cdot}09 \approx 6{\cdot}9 \text{ years}$$

Superimposing these results on Figure 5.18 we find that they fall exactly onto the previously plotted curve.

5.3.3. DETERMINATION OF THE COEFFICIENT OF CONSOLIDATION

In the previous section the coefficient of consolidation, c_v, was determined from knowledge of the coefficient of permeability, k, and the coefficient of volume compressibility, m_v ($m_v = 1/E_{oed}$). The determination of these two coefficients requires two separate tests and, although, in certain cases, this may lead to more accurate results, the possibility of determination of this coefficient directly from the oedometer test must be investigated. This, in fact, can be achieved with the help of the theory of consolidation.

Let us assume that under a given load increment in the oedometer the experimental time to achieve a certain degree of consolidation is known (e.g. t_{50} for $U_0 = 0{\cdot}5$). If this is the case, then we can utilize Equation (5.36) to determine c_v:

$$c_v = \frac{4h^2}{\pi^2 t_{50}} N_0$$

where, from Table 5.5, $N_0 = 0{\cdot}49$ for $U_c = 0{\cdot}50$. The problem, therefore, reduces to the determination of the experimental time required to achieve a known degree of consolidation.

The direct comparison between experimental and theoretical time–consolidation curves is complicated by the presence of secondary consolidation and hence empirical methods must be used. These are called curve-fitting methods. The two methods recommended by British Standard 1377 (1967) (Test No. 16) are as follows.

(a) 'Square root of time' fitting method in which the experimental time corresponding to 90% consolidation is determined ($c_v = 0{\cdot}848h^2/t_{90}$),

(b) 'Logarithm of time' fitting method in which the experimental time corresponding to 50% consolidation is determined ($c_v = 0{\cdot}196h^2/t_{50}$).

A brief description of the second method is given below.

The initial height of the sample at the beginning of the consolidation process is determined according to the construction shown in Figure 5.20(b), where $\sqrt{t}$ (in minutes) is plotted against the height of the sample. The intersection of a straight line through the experimental points with the vertical axis is taken as the initial height ($U_0 = 0$).

The consolidation curve plotted to a semilogarithmic scale (Figure 5.20(a)) is usually convex upwards in the initial stages of the consolidation, concave in the latter stages, and finally changes to a straight line inclined to the horizontal axis. By plotting two tangents to this curve, one at the point of contraflexure and the other to the final straight part of the curve, a point of intersection is obtained which is taken as the end of primary consolidation (due to expulsion of water)

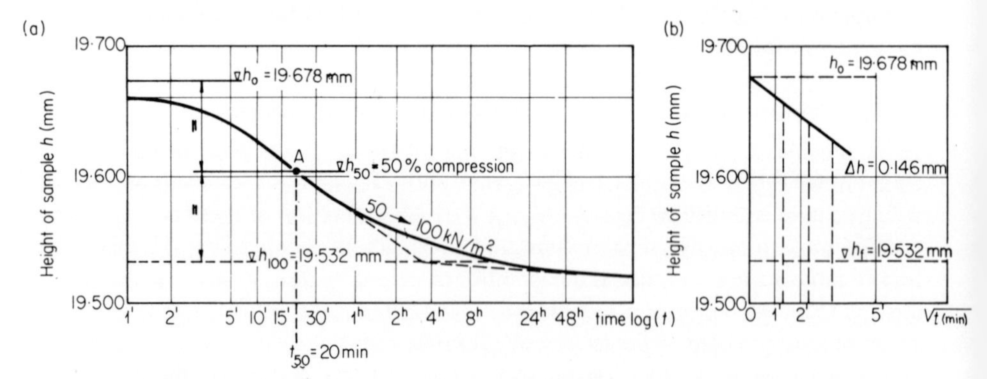

Figure 5.20. Details of logarithm of time-fitting method; (a) time–consolidation curve; (b) origin correction.

(Lambé, 1951). Thus the height of the sample corresponding to $U_c = 1{\cdot}0$ is obtained. The 50% compression point is half-way between h_0 and h_{100}, i.e. at A. The time, t_{50}, corresponding to $U_c = 0{\cdot}5$ can now be read off the horizontal axis and used in evaluation of the coefficient of consolidation ($c_v = 0{\cdot}196 h_0^2/t_{50}$).

5.3.4. SECONDARY CONSOLIDATION

Compression beyond the empirically established end of primary consolidation (Figure 5.20(a)) is referred to as *secondary consolidation* or *creep* and is not dependent on the expulsion of free water from the soil. It is probably due to some slippage between particles and to a very slow expulsion of some of the strongly held water molecules from the adsorbed water (at the point of contact between clay particles) to the free water, i.e. dehydration of the adsorbed cations.

Because of the strong influence of test conditions on the rate of secondary consolidation, extrapolations from laboratory data to estimate field behaviour are not very reliable at present. With the exception of consolidation of organic soils or very sensitive clays, the effects of secondary consolidation can usually be disregarded.

5.3.5. SOURCES OF ERROR IN ESTIMATION OF RATES OF DEFORMATION

The oedometer test, on the basis of which estimates of the rates of settlement are made, is basically a model test and therefore one cannot expect accurate results from it unless all the conditions of similarity between the laboratory and field conditions are satisfied. A large proportion of natural soils are non-homogeneous. Some are laminated and some may contain a network of open joints through which water can move with much greater ease than through the intact lumps between them; in such cases the process of consolidation will not even approximate to the one-dimensional process described in the preceding section. To deal with these cases, which are outside the scope of this book, the theory of consolidation has been extended to two and three dimensions and new laboratory and *in situ* tests have been developed to obtain the necessary soil parameters (see, for example, Lambé and Whitman, 1969).

5.4. Strength of Soils

5.4.1. SHEAR STRENGTH OF SOILS

The shear strength of a soil is the maximum available resistance that it can offer to shear stress at a given point within itself. When this resistance is reached continuous shear displacement takes place between two parts of the soil body.

Shear displacements can take place across a well-defined single rupture plane or across a wide shear zone (containing a zone of soil stressed to its available resistance). In both cases the condition for generation of continuous shear deformations (sliding) is the mobilization of the available shear strength of soil by the acting shear stress:

$$|\tau| = \tau_f \tag{5.37}$$

where $|\tau|$ = absolute value of shear stress
τ_f = shear strength of soil

Stress σ acting at any point within the soil mass (Figure 5.21) can be resolved into two components:

σ_n, stress component normal to the plane in which the equilibrium conditions are being considered

τ, shear stress within the same plane

Shear strength of the soil τ_f is mobilized within the plane of action of shear stress τ, but is acting in the opposite direction. The magnitude of the resistance is evaluated from Coulomb's equation:

$$\tau_f = c + \sigma_n \tan \phi \tag{5.38}$$

where $\sigma_n \tan \phi$ = internal frictional resistance in kN/m^2
σ_n = normal stress component (perpendicular to the shear plane) in kN/m^2
ϕ = angle of internal friction in degrees
c = cohesion (cohesive resistance) in kN/m^2

For cohesionless soils Equation (5.38) simplifies to

$$\tau_f = \sigma_n \tan \phi \tag{5.39}$$

because cohesion $c = 0$.

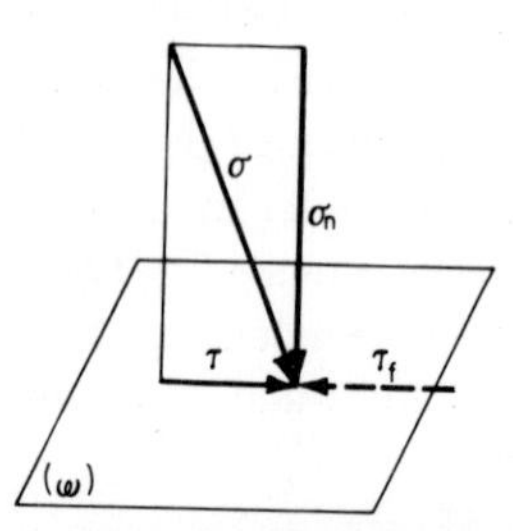

Figure 5.21. Schematic illustration of shear plane stresses.

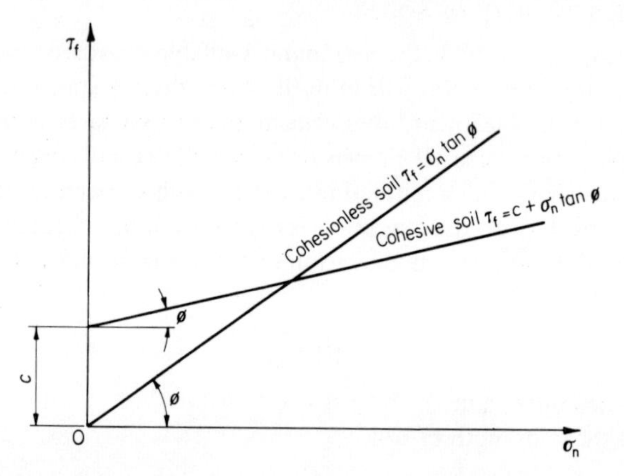

Figure 5.22. Shear strength of soil as a function of normal stress and Coulomb's parameters ϕ and c.

Graphical representations of Equations (5.38) and (5.39) are shown in Figure 5.22.

5.4.2. ANGLE OF INTERNAL FRICTION (SHEARING)

The phenomenon of sliding frictional resistance is well known in the engineering science. In the case of shearing of granular soils we are dealing with a combination of rolling and sliding friction, because in shear displacement of one layer of soil over another, across a wide shear zone, the resistance is not only due to the friction within the sliding planes but also due to a relative rolling movement between neighbouring grains (Figure 5.23).

As can be seen in the diagram all the grains within the shear zone are rotating in the same direction which results in opposite relative movement

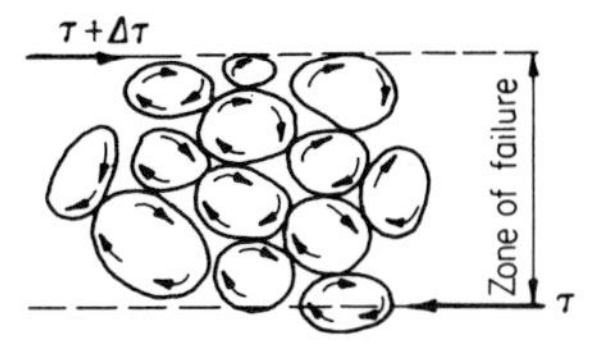

Figure 5.23. Rotation of soil grains within the shear zone.

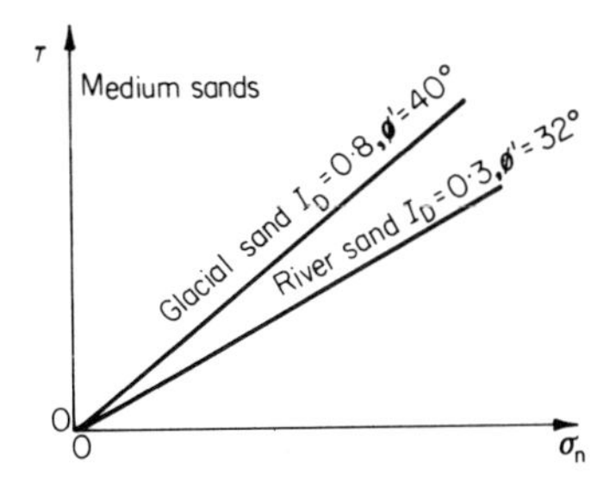

Figure 5.24. Variation of ϕ' with the degree of compaction and shape of sand grains.

between their surfaces, and hence leads to generation of friction at the points of contact between them.

The resistance of soil due to the sliding and rolling friction is known as the true *internal frictional resistance.*

The internal shearing resistance of soil is built up of the true *internal frictional resistance* and of the resistance which the soil possesses by virtue of the *interlocking* of individual grains. Interlocking contributes a large proportion of the strength in dense (overconsolidated) soils while it has little or no effect on the strength of very loose (normally consolidated) soils; the gradual loss of strength after the peak point is passed, illustrated by results of Test 3.14 in Figure 5.26 may be attributed to a gradual decrease in interlocking which takes place because the sample is decreasing in density (Taylor, 1948).

The internal shearing resistance is a function of the interparticle (effective) stress, normal to the plane of shearing and the angle of internal friction (it is also referred to as the angle of internal shearing resistance).

The *angle of internal friction*, in spite of its name, does not depend solely on the internal friction between grains and particles because a portion of the shearing stress on the plane of failure is utilized in overcoming interlocking, i.e. it is also dependent on the initial voids ratio or density of a given soil (Figure 5.24). The magnitude of the angle of internal friction also depends on the grain size and on their shape: the larger the grains are, the wider is the zone affected by the internal friction; and, the more angular they are, the greater is the frictional resistance to their relative movement. The extent of influence of the above factors on the angle of internal friction of sands is illustrated in Figure 5.25 (Rowe, 1962).

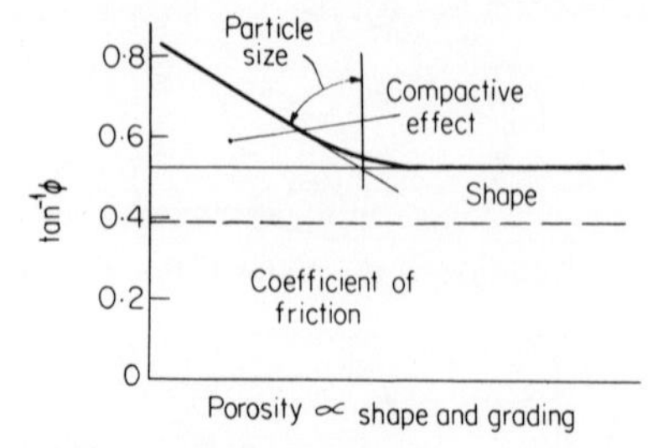

Figure 5.25. Build-up of the angle of internal friction in granular soils (after Mackey, 1964; by permission of the Editor of *Civil Engineering* and *Public Works Review*).

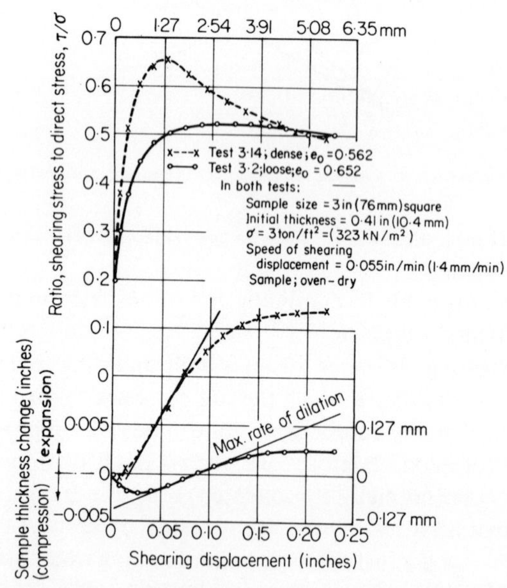

Figure 5.26. Results of typical direct shear tests on Ottawa standard sand (after Taylor, 1948; by permission of John Wiley and Sons Ltd.).

Shearing of soils is generally accompanied by volume changes (Figure 5.26); dense soils increase in volume or dilate, whereas very loose soils decrease in volume or compress. After considerable straining of any soil both the shearing resistance and the voids ratio become constant (Figure 5.26). This condition is referred to as the critical or constant volume condition. The angle of internal friction corresponding to that condition is usually referred to as the *critical angle of friction* and the voids *ratio as the critical voids ratio*. For any given soil both these parameters are independent of its initial voids ratio (density) but are a function of the normal effective stress at which the shearing occurs. In the case of clays of both normally consolidated and overconsolidated type there is a further loss of internal shearing resistance due to a gradual reorientation of clay particles into parallel face-to-face arrangements. This seems to take place after the clay has reached essentially constant volume, i.e. critical voids ratio. The angle of friction corresponding to this ultimate (residual) state is referred to as the *residual angle of friction* which for soils with clay content approaching 100% is of the same magnitude as the angle of friction between clay crystals (which can be as low as 3° to 4°), whereas for soils with low clay content it approaches that between quartz grains, i.e. about 32° (Skempton, 1964).

The angle of internal friction of submerged sands is only slightly different from that obtained for the same sands in perfectly dry condition.

Presence of clay fraction in granular soil will reduce its internal friction because clay particles tend to coat the larger grains and assist in interparticle sliding.

As can be seen from the above considerations the angle of internal friction of any soil is not a unique mechanical property but it can vary between the *initial* (*peak*) value and the *ultimate* or *residual* value.

5.4.3. COHESION

In a body of a cohesive soil the particles are in the state of static equilibrium. To bring them closer together it is necessary to apply an external compressive force; to pull them apart it is necessary to apply an external tensile force; to displace them tangentially it is necessary to overcome the resistance offered by the existing bonds between the particles, i.e. by the cohesion.

True cohesion in soils is the resistance to shearing that they offer as the result of the existence of molecular (electro-chemical) attractive forces between their small particles (see Section 2.6).

In analysis of factors which govern the magnitude of the true cohesion it can be assumed that for any soil of a given granulometric and mineralogical composition its magnitude depends on the number of interparticle contacts per unit area of the shear plane and on the distance between the particles; therefore it can be said that the true cohesion depends on the number of particles per unit volume of a given soil.

The number of particles per unit volume of a given soil depends in turn on the water content; the greater the number of particles of a given soil per unit volume, the smaller is its water content and hence the greater should be its true cohesion. The above deductions have been confirmed by experimental evidence (Figure 5.27).

The influence of the type of soil on the true cohesion can be summarized as follows: loose granular soils, not containing any clay particles, have the true cohesion equal to zero; as the clay content in a soil increases so does its true cohesion, because the number of particles per unit volume increases, i.e. the number of interparticle contacts per unit area of a shear plane increases. Because, however, the molecular attractive forces depend to a large extent on the mineralogical composition of the particles and on the type and concentration of electrolytes (present in the pore water), the magnitude of the true

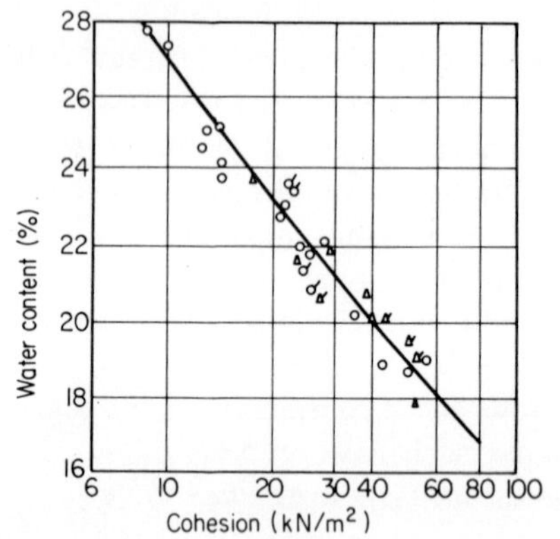

Figure 5.27. Relationship between cohesion and water content (after Bjerrum, 1954).

cohesion depends on these factors as well as on the number of clay particles present.

The cohesion of natural soils can be greater than their true cohesion (as obtained for the soil remoulded at the natural water content) because of the presence of diagenetic bonds or cementation of particles due to crystallization of salts in the pores. However, these apparent increases in the cohesion cannot always be taken into consideration because they are not permanent and can easily be destroyed by weathering or upon immersion in water.

5.4.4. SHEAR STRENGTH OF COHESIVE SOILS

Shear strength of cohesive soils is taken as the combined resistance of the internal frictional resistance and the cohesion as given by the Coulomb–Hvorslev equation:

$$\tau_f = \sigma'_n \tan \phi'_e + c'_e \tag{5.40}$$

where τ_f = shear strength of soil in kN/m²
σ'_n = normal effective stress in kN/m²
ϕ'_e = true angle of internal friction in degrees
c'_e = true cohesion in kN/m²

The basic strength parameters c'_e and ϕ'_e* are difficult to measure for many soils and have a limited direct application in analyses of practical problems. At the present time therefore Coulomb's equation, Equation (5.38), is generally used:

$$\tau_f = \sigma_n \tan\phi + c$$

in which the values of strength parameters c and ϕ are determined from tests under definite conditions of water content change (drainage) during shearing which, as closely as possible, resemble the field working conditions of the given soil.

In the case of very permeable soils or in the case of slow rates of loading of soils of low permeability, such that there is no increase in the pore water pressure, i.e. $\Delta u = 0$, the whole load increment is taken up by the soil skeleton; the effective normal stress is equal to the total normal stress $\sigma'_n = \sigma_n$ (Figure 5.28(a)) and full water content change takes place. Such practical

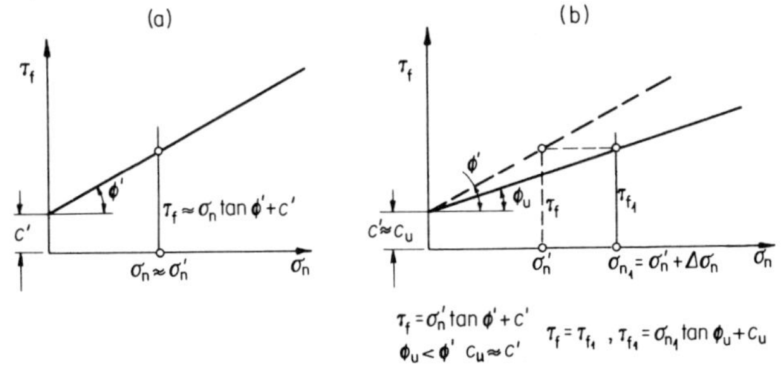

Figure 5.28. Relationship between shear strength and normal effective and total stresses: (a) slow loading with full dissipation of pore water pressure (drained); (b) quick loading with build-up of pore water pressure (undrained).

conditions are referred to as fully drained and strength parameters obtained under these conditions, i.e. with respect to the effective stresses, are denoted by c' and ϕ' (or c_d, ϕ_d) and are referred to as drained parameters.

In the case of quick application of loading to saturated soils of low permeability there is no immediate change in water content and hence pore water pressure increases in parallel with the increase in the applied loading $\Delta u \approx \Delta\sigma_n$ and therefore the effective stress remains constant: $\sigma_{n_1}' = \sigma'_n$ (Figure 5.28(b)). In consequence, in spite of an increase in the total normal stress from σ_n to $\sigma_{n_1} = \sigma_n + \Delta\sigma_n$, there is no increase in the shearing resistance of the soil ($\tau_{f_1} = \tau_f$) and hence the new angle of internal friction ϕ_u obtained with respect to total stresses is smaller than ϕ'. Such practical conditions are referred to as undrained and strength parameters c_u and ϕ_u are obtained with respect to the total stresses.

* Parameters c'_e and ϕ'_e define the shear strength of a given soil only for a particular constant water content (voids ratio) at the instant of shear failure and analyses based on these properties are at present too elaborate for practical use.

Results of laboratory tests shown in Figure 5.29 confirm the above arguments and also demonstrate a marked increase in the shear strength due to consolidation of the soil prior to shearing.

It follows from the above observation that under quick or undrained loading and shearing of cohesive soils the angle of internal friction ϕ_u is smaller than the drained angle of friction ϕ' and that their cohesion is increased by consolidation prior to shearing. For practical purposes the above can be summarized as follows.

(1) Under fully drained conditions (slow rate of loading) cohesive soils consolidate and their shearing resistance increases owing to the increase in the interparticle (effective) stresses and owing to the densification of the soil.

(2) In undrained conditions (quick loading) cohesive soils, sheared under the same normal stress σ_1 have a much smaller shearing resistance (in relation to drained conditions) because there is no increase in the effective stresses and no change in the packing of particles (constant volume).

British Standard Code of Practice CP 2004 recommends, in line with practices in other countries, that shearing resistance of soil should be determined under conditions under which the material will be stressed in the field. Laboratory tests which satisfy most of the field conditions are described in British Standard CP 2001 (1957, Appendix H), and may be classified as follows.

(a) Undrained (quick) test: for structures in which the live load applied after completion of construction is greater than 70% of the total load the effects of consolidation (during the construction period) on the strength of non-fissured cohesive soils can be neglected and the samples are sheared in an undrained manner; results are plotted in terms of total stresses and strength parameters c_u and ϕ_u are obtained.

(b) Consolidated undrained (consolidated quick) test: for structures in which the live load represents between 30 and 70% of the total load a partial increase in the strength of cohesive soils due to consolidation during construction can be considered. Samples are consolidated under stress conditions varying between the *in situ* (overburden) and that corresponding to the dead weight of the structure and are then sheared in an undrained manner. Again, strength parameters c_u and ϕ_u are obtained.

If facilities are available for pore water pressure measurement during the undrained stage of the test, then the same procedure can be used for determination of the effective strength parameters c' and ϕ'. These, together with *in situ* measurements of the pore water pressures can be used for stability analysis of cohesive soils under slowly increasing loading, e.g. under slowly constructed embankments.

(c) Drained (slow) test: if the live load represents less than 30% of the total load and the period of construction is sufficiently long to allow full water content changes to take place (which in the case of slightly cohesive soils corresponds to the normal rate of construction), full increase in the strength due

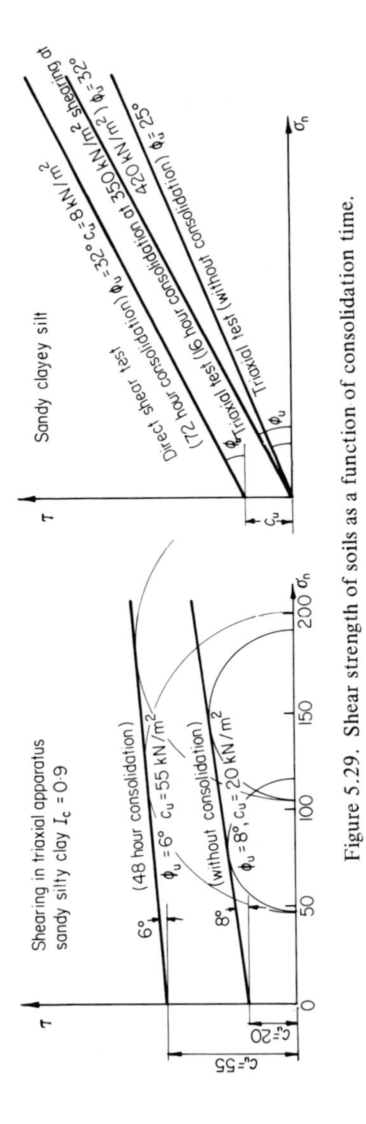

Figure 5.29. Shear strength of soils as a function of consolidation time.

to consolidation can be considered. The samples are consolidated under the same stress conditions as in the case of the consolidated undrained test but are then sheared in a drained manner, i.e. at such a slow rate that there is no increase in the pore water pressure; the effective strength parameters c' and ϕ' are obtained.

In determination of the effective strength parameters for the analysis of long-term stability of slopes in cuttings the samples are softened under stress conditions varying between initial *in situ* and final *in situ*, and are then sheared in a fully drained or undrained (with pore water pressure measurement) manner; the results are plotted in terms of effective stresses.

Typical values of strength parameters of cohesive soils are given in Table 5.7 (after the Polish Code of Practice PN-59B-03020, 1959) and it is suggested that in the absence of laboratory test results these values can be used in evaluation of safe bearing stresses. It must be emphasized, however, that in order to be able to use these values properly it is necessary to classify the soils correctly and to determine their physical properties.

5.4.5. LABORATORY METHODS OF DETERMINATION OF COHESION AND ANGLE OF INTERNAL FRICTION

Laboratory shear tests are usually carried out in a direct shear apparatus (shear box) or in a triaxial compression apparatus. For certain soils simple unconfined compression apparatus or laboratory shear vane can be used.

The choice of apparatus is primarily determined by the condition of drainage under which it is desired to carry out the test but is also influenced by the type of the soil to be tested. Table 5.8 (Skempton and Bishop, 1950) summarizes the procedures generally followed in testing the main soil types. Tests are normally carried out under a controlled rate of strain* as the conditions at failure are then completely specified.

In direct *shear apparatus* (Figure 5.30) the soil sample, usually 60 mm square by 20 mm thick or 12 inches (305 mm) square by 6 inches (102·5 mm) thick (large shear box), is placed in a horizontally split box and after consolidation under a predetermined force Q is sheared by application of a horizontal force T.

From the results of several shearing tests, under different normal stresses (within the range of the working field stresses), Coulomb's failure envelope can be plotted as shown in Figure 5.31; the intercept on the shear stress axis gives the cohesive resistance c and the slope of the line gives the angle of the internal shearing resistance ϕ.

The use of a shear box is normally limited to drained tests, because, owing to the small thickness of the sample (short drainage path), undrained and

* For the undrained (quick) test British Standard 1377 requires failure to take place within a period of approximately 5 to 10 minutes.

Table 5.7. Typical values of strength parameters ϕ and c in kN/m² (after Polish Code PN-59/B-03020, 1959)

		Type of soil		Density index of cohesionless soils			
			I_D=	1·0 – 0·67	0·67 – 0·33	0·33 – 0	
cohesionless	inorganic	gravels, tills, hoggins, etc.	ϕ'	45°–40°	40°–37°	37°–35°	
cohesionless	inorganic	sands: coarse and medium	ϕ'	40°–38°	38°–35°	35°–32°	
cohesionless	inorganic	sands: fine and silty	ϕ'	37°–35°	35°–32°	32°–28°	
cohesionless	organic	sands, organic	ϕ'	30°–25°	25°–22°	22°–18°	
				Consistency of cohesive soils			
				hard or very stiff	stiff	firm	soft to very soft
				$w = w_s$ I_c = 1·0 – 0·75	0·75 – 0·50	0·50 – 0·0	
cohesive	inorganic	slightly clayey sands, sandy silts, silts $J < 10\%$	ϕ' c' ϕ_u	28°–24° 40*–30 25°–20°	24°–22° 30–20 20°–16°	22°–19° 20–15 16°–10°	19°–5° 15–2 10°–7°
cohesive	inorganic	clayey sands, clayey sandy silts, clayey silts, $J = 10–20\%$	ϕ' c' ϕ_u	26°–22° 50–40 20°–16°	22°–19° 40–30 16°–12°	19°–15° 30–20 12°–7°	15°–12° 20–3 7°–5°
cohesive	inorganic	sand–clays, sand–silt–clays, silt–clays $J = 20–30\%$	ϕ' c' ϕ_u	23°–20° 60–50 15°–12°	20°–17° 50–40 12°–9°	17°–12° 40–30 9°–5°	12°–8° 30–5 5°–2°
cohesive	inorganic	sandy clays, clays, silty clays $J > 30\%$	ϕ' c' ϕ_u	19°–17° 80–60 10°–8°	17°–14° 60–50 8°–5°	14°–5° 50–40 5°–2°	10°–5° 40–10 2°–0°
cohesive	organic	organic silts, peats, etc.		all strength parameters to be determined from laboratory tests			

* For approximate conversion from kN/m² to lbf/ft² multiply by 21 and to kgf/cm² by 0·01.
For computation of safe bearing capacity undrained cohesive resistance c_u can be taken as equal to c' – this assumption is on the safe side.

Table 5.8. Choice of apparatus for determination of shear strength for main soil types (after Skempton and Bishop, 1950)

	Soil type	Test conditions		
		Undrained	Consolidated undrained	Drained
cohesionless	gravels	—	—	large shear box
cohesionless	sands	—	triaxial	triaxial or shear box
cohesionless	clay–gravel	triaxial or large shear box (if of low permeability)	large shear box or triaxial depending on grading and permeability	large shear box or triaxial depending on grading
cohesionless	silts	triaxial	triaxial	triaxial or shear box
cohesive	non-fissured clays	triaxial or unconfined compression or shear vane	triaxial (shear box) or oedometer + triaxial	triaxial or shear box
cohesive	fissured clays	triaxial	triaxial or oedometer + triaxial	triaxial
organic	organic soils, peats, etc.	triaxial	triaxial	triaxial or shear box

consolidated undrained tests can only be carried out on clays of low permeability at high testing rates.

Example. Four 60 mm^2 square by 20 mm thick samples of the same soil were sheared in a direct shear apparatus under vertical loads of 36·7, 55·0, 73·4, and 91·7 kg. The corresponding measured maximum horizontal shearing forces were 252, 346, 425, and 540 N. Determine the cohesive resistance and the angle of internal friction.

Sample No.	Q (kg)	σ'_n (kN/m^2)	T (N)	τ_f (kN/m^2)
1	36·7	100	252	70
2	55·0	150	346	96
3	73·4	200	425	118
4	91·7	250	540	150

With shear stress τ_f as ordinate and normal stress σ'_n as abcissa plot the results (Figure 5.31); a straight line through the given points determines $c' = 20$ kN/m^2 and $\phi_u = 27°$.

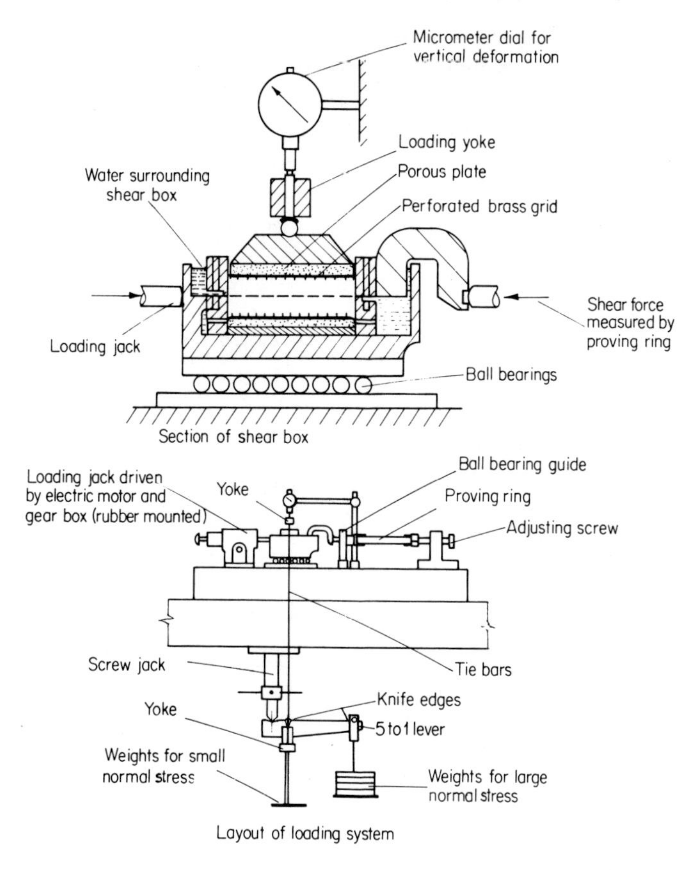

Figure 5.30. Schematic diagram of direct shear apparatus (shear box) (after Skempton and Bishop, 1950; by permission of the Council of the I.C.E.).

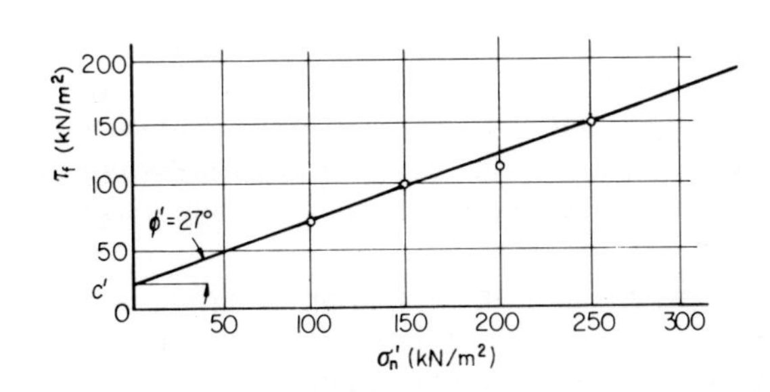

Figure 5.31. Graphical determination of ϕ' and c' from shear box test results.

Shear test in *triaxial compression apparatus* (Figure 5.32) is also carried out on several (at least three) samples of the same soil.

The test specimens are in the form of right cylinders of nominal diameter between 38 mm (1½ in) and 110 mm (4½ in) and of height approximately equal to twice the nominal diameter (British Standard 1377, 1967). A thin rubber membrane is placed over the sample prior to testing, isolating it from the cell water, and connecting it with the bottom and top drainage filters. The cell is

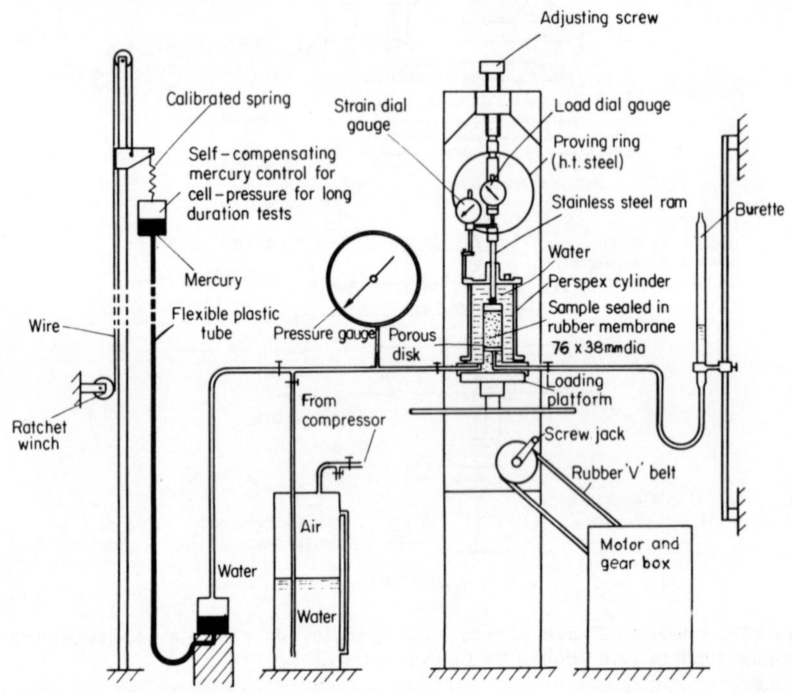

Figure 5.32. Diagrammatic layout of triaxial apparatus for testing saturated or partly saturated soils (after Skempton and Bishop, 1950; by permission of the Council of the I.C.E.).

now placed over the sample and is filled with water which is pressurized to the required working pressure.

Shearing is now carried out by applying additional vertical stress q which increases until the shearing resistance of the sample has been overcome; the maximum sum total of the cell pressure σ_3 and the additional stress q is denoted by σ_1.

As there are no external shear stresses acting on the sample σ_1 and σ_3 are the principal stresses; the test specimen is sheared along a plane inclined at angle α to the horizontal (Figure 5.33), and the magnitude of normal stress σ_n and shear stress τ acting on that plane are determined from Mohr's circle.

The obtained Mohr's circle is a limiting circle defining the stress conditions at failure and the measured shear stress τ is, for the given σ_n, equal to the shear strength τ_f of the soil.

From a number of tests on specimens of the same soil, at different pressures σ_3, a series of limiting Mohr's circles is obtained. A line tangential to these circles is obviously the Coulomb failure envelope (Figure 5.34) whose intercept with the ordinate axis gives the cohesion c and whose slope is equal to the angle of internal friction ϕ.

In order to be able to observe clearly the actual moment of failure of specimen tested in triaxial compression the shearing is carried out at a constant rate of strain. This enables the determination of the maximum deviatoric stress q.

According to the British Standard 1377 (1967) the cross-sectional area of the sample used for evaluation of q_{max} is modified to allow for the lateral strains

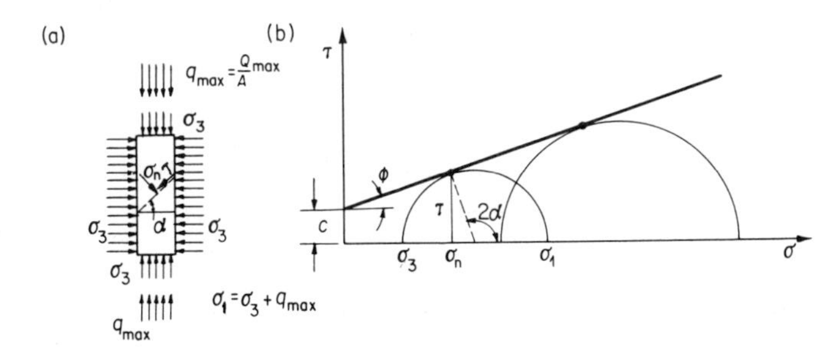

Figure 5.33. Determination of stresses in sample tested in triaxial compression apparatus: (a) stresses acting on test specimen; (b) determination of σ_n and τ from the limiting Mohr circle.

which take place during shearing: in undrained shearing of saturated soils it is assumed that deformations occur at constant volume, i.e. Poisson's ratio is taken as equal to 0·5, whereas in drained shearing the actual volume changes are measured which, together with the observed vertical strains, enable the necessary correction to be made to the cross-sectional area.

Example. A cylindrical specimen of saturated soil is being sheared in triaxial apparatus in an undrained manner and, at a given time t, its vertical compression is equal to $\Delta h = 6{\cdot}08$ mm. Assuming that its initial height and cross-sectional area were respectively $h_0 = 76{\cdot}0$ mm and $A_0 = 1138$ mm^2, determine its cross-sectional area at the time t.

Let the vertical strain at the time t be $\epsilon_1 = -\Delta h/h_0$ (compressive strain is taken as negative) and the lateral strains $\epsilon_2 = \epsilon_3$. For no volume change the volumetric strain $\epsilon_v = 0$. From definition of volumetric strain $\epsilon_v = \epsilon_1 + \epsilon_2 + \epsilon_3$ and for radial symmetry $\epsilon_v = \epsilon_1 + 2\epsilon_3$. Hence, for $\epsilon_v = 0$, $2\epsilon_3 = -\epsilon_1 = \Delta h/h_0 = 0{\cdot}08$. The increase in the cross-sectional area $\Delta A \approx 2\epsilon_3 A_0$ and thus the new area A_t is

$$A_t = A_0 + \Delta A = A_0(1 + 2\epsilon_3) = A_0(1 - \epsilon_1)^*$$
$$= A_0\left(1 + \frac{\Delta h}{h_0}\right) = 1223 \text{ mm}^2$$

Example. On shearing in triaxial compression of three specimens of the same soil the following values of σ_3 and corrected σ_1, were obtained.

	Sample 1	Sample 2	Sample 3
σ_3 (kN/m^2)	50	150	250
σ_1 (kN/m^2)	120	250	370

The angle of internal friction ϕ_u and the cohesive resistance c_u are obtained by plotting the limiting Mohr stress circles in Figure 5.34. The common tangent

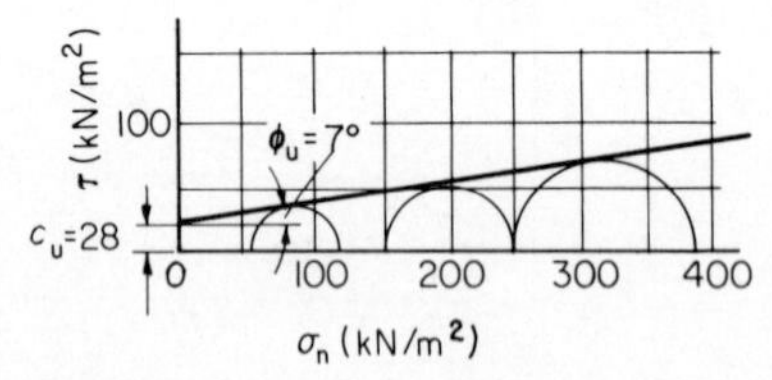

Figure 5.34. Graphical determination of ϕ_u and c_u from triaxial compression test results.

(failure envelope) to the Mohr circles gives c_u = 28 kN/m^2 (39·8 lbf/in^2, 0·28 kg/cm^2) and $\phi_u = 7°$.

5.4.6. REVIEW OF DIFFERENT METHODS OF SHEARING

The determination of shear strength of soils is one of the most important and at the same time one of the most complex problems in soil mechanics. This is mainly due to the lack of understanding of the role of the interparticle forces within the soil skeleton. The solution to this basic problem must be sought within the realm of the physical chemistry of granular media.

The present engineering solutions to these problems are, to a large extent, based on a good engineering judgement or intuition and on detailed observations of the manner of behaviour of soils at failure; they are therefore approximate, but frequently sufficiently accurate, for practical purposes.

The first hypothesis of the shear strength of soils was presented by Coulomb in 1773 in the well-known form (Equation (5.38))

$$\tau_f = c + \sigma_n \tan \phi$$

* In a drained test $\epsilon_v \neq 0$ and hence $2\epsilon_3 = \epsilon_v - \epsilon_1$ and $A_t = A_0(1 + \epsilon_v - \epsilon_1) = A_0(1 + \Delta V/V_0 + \Delta h/h_0)$, where ΔV is the measured change in volume (negative if sample decreases in volume) and V_0 is the initial volume of the specimen.

However, as is now well known, the strength parameters c and ϕ are not constants but depend foremost on the actual water content of the soil. As early as 1925 Terzaghi had pointed out the necessity of consideration of the effects of changes in internal forces and in water content on the strength of the soil.

5.4.6.1. Krey–Tiedemann Method (*Horn, 1964*)

The first experiments to investigate these effects were carried out in 1927 by Krey. In his tests, which were carried out in a specially constructed direct shear apparatus, remoulded clay samples were initially consolidated under different normal stresses σ_k and were then sheared immediately after the normal stresses were reduced to σ_n. Similar experiments were carried out by Tiedemann who then suggested the following expression for the shear strength of remoulded clays:

$$\tau_f = \sigma_k \tan \phi_c + \sigma_n \tan \phi_K \tag{5.41}$$

i.e. suggesting that

$$c_K = \sigma_k \tan \phi_c \tag{5.42}$$

Schematic interpretation of Krey–Tiedemann experimental results are shown in Figure 5.35, from which it can be seen that they have obtained a definite relationship between the water content of the consolidated soil and its shear strength. The very important achievement of these two research workers was the

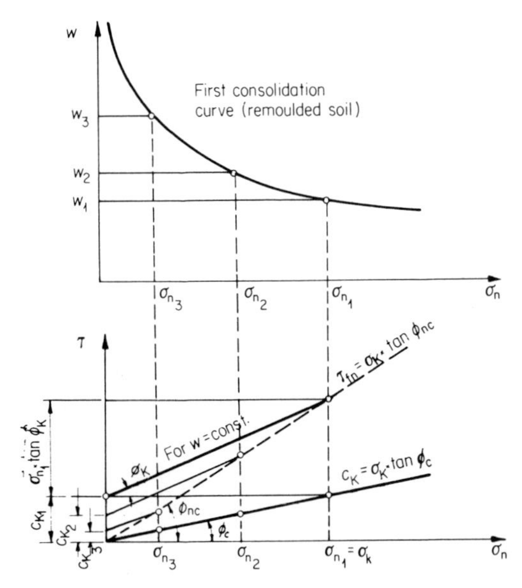

Figure 5.35. Experimental results according to Krey–Tiedemann method.

determination of the main branch of the shear strength envelope which defines the strength of a normally consolidated soil, i.e. of a soil which has been sheared under a normal stress σ_n equal to the consolidation stress σ_k:

$$\tau_{fn} = \sigma_k \tan \phi_{nc} \tag{5.43}$$

The main branch of the strength envelope passes through the origin at an angle ϕ_{nc} to the abscissa (Figure 5.35).

The second important achievement was the determination of the relationship

$$c_K = \sigma_k \tan \phi_c \tag{5.44}$$

Krey–Tiedemann's work has proved beyond doubt that the shear strength of soils is dependent on the consolidation stress and on the water content. However, the obtained values of c_K are rather high because no swelling was permitted prior to shearing which would, undoubtedly, result in the reduction of τ_f.

Experimental results obtained by Wiłun (Figure 5.36) demonstrate that as the swelling progresses the value of c_K decreases and thus the shear strength of

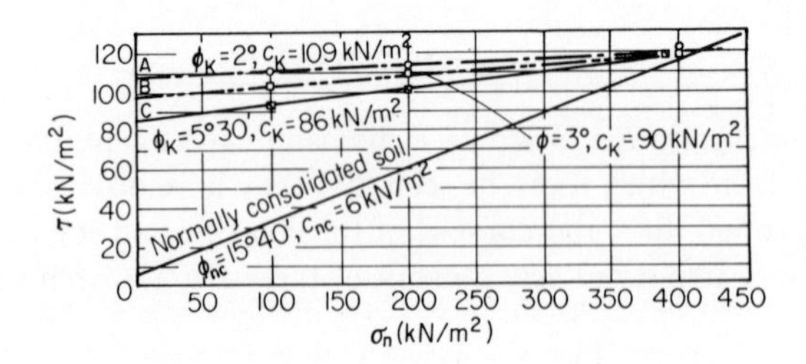

Figure 5.36. Decrease in cohesion c_K with relation to deconsolidation time: A, deconsolidation time, 3 minutes; B, 15 minutes; C, 24 hours.

the soil is reduced. This is particularly important in the analysis of stability of slopes in cuttings where, because of the reduction in the overburden stresses, shear strength of the soil decreases owing to expansion.

5.4.6.2. Terzaghi–Hvorslev Method (*Hvorslev, 1936*)

In 1931 Terzaghi postulated a new shear strength hypothesis in which he related the shear strength to the effective (interparticle) stresses and introduced new concepts of the true cohesion c'_e and true angle of internal friction ϕ'_e acting over the contact areas between particles. Terzaghi's hypothesis was experimentally verified by Hvorslev in 1934–7 who also explained the nature of the hysteresis loop observed during the consolidation, deconsolidation, and reconsolidation of soils and of its equivalent loop in the shear strength envelope obtained by shearing of remoulded soils at various stages of the above loading cycle. Using triaxial apparatus he has carried out a series of undrained shear tests on remoulded clay samples in which pore water pressures were measured.

The resultant Terzaghi–Hvorslev equation

$$\tau_f = c'_e + \sigma'_n \tan \phi'_e \tag{5.45}$$

is based on consideration of effective stress $\sigma'_n = \sigma_n - u$ and true strength parameters c'_e and ϕ'_e (see Sections 5.4.2 to 5.4.4). Schematic interpretation of the Hvorslev results and the interdependance between the consolidation and shear strength diagrams are shown in Figure 5.37.

Undoubtedly the Terzaghi–Hvorslev cohesion intercepts c'_e are considerably smaller than those obtained by Krey–Tiedemann (c_K) and conversely ϕ'_e is greater than ϕ_K. Values of c'_e and ϕ'_e can be considered as constants for a given

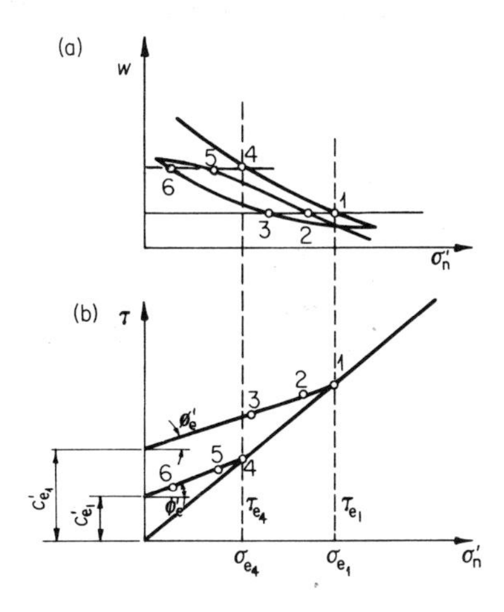

Figure 5.37. Experimental results according to Terzaghi–Hvorslev method: (a) water content–effective stress consolidation curve; (b) shear strength envelope.

soil, the former being a function of the water content only. The shear strength determined from Equation (5.45) is the true resistance that a given soil will offer to shear deformation.

It must be emphasized that the true strength parameters c'_e and ϕ'_e define the shear strength of a given soil at a specific constant water content corresponding to the consolidation stress σ_e (Figure 5.37(a)).

In practice, however, we have to deal with variable water contents dependent on the actual effective stresses. This problem has been solved by Skempton and Bishop and is dealt with in the following section.

5.4.6.3. Skempton–Bishop Method

For nearly three decades now the problem of the pore water pressures has been one of the major topics of research and contributions in this field of Skempton and Bishop are of particular importance. They have introduced two basic shear strength equations:

for normally consolidated soils: $\tau_f = (\sigma - u) \tan \phi'$ (5.46)

for overconsolidated soils: $\tau_f = c' + (\sigma - u) \tan \phi'$ (5.47)

where c' denotes the apparent cohesion } in terms of
ϕ' denotes the angle of internal shearing resistance } effective stresses
σ denotes the total stress normal to the plane considered
u denotes the pore water pressure

The shear strength envelope for normally consolidated soils, when plotted in terms of effective stresses, is a straight line which passes through the origin.

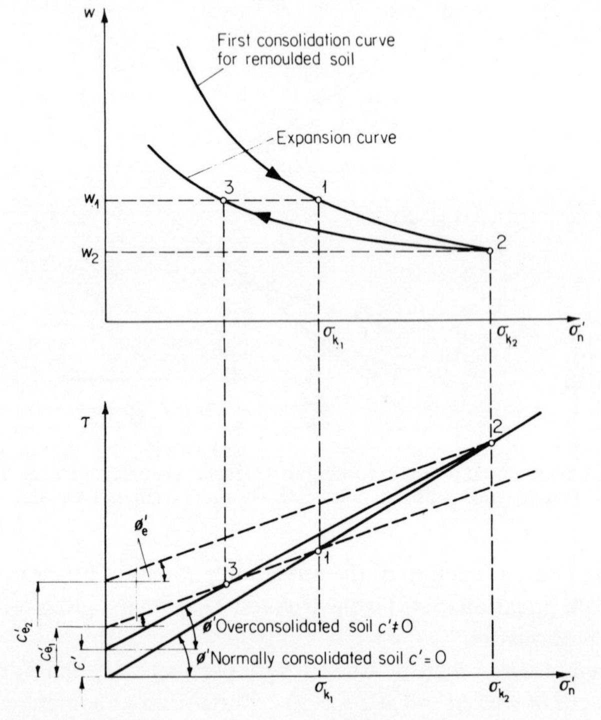

Figure 5.38. Comparison of test results and of the methods of interpretation according to Terzaghi–Hvorslev and Skempton and Bishop.

On shearing of overconsolidated soils a slightly curved failure envelope is obtained, which for the range of stresses under consideration can be approximated by a straight line with a cohesion intercept c' (the value of which is a function of the consolidation pressure σ_k) and slope ϕ' as shown in Figure 5.38. The values of c' are smaller than c_K and c'_e while ϕ' is greater than ϕ'_e. The relationship between c'_e, ϕ'_e and c', ϕ' is illustrated in Figure 5.38.

According to Skempton and Bishop the strength parameters c' and ϕ' can be determined either from consolidated drained tests (in which case they are frequently denoted by c_d and ϕ_d) or from consolidated undrained tests with pore water pressure measurements. The latter method requires triaxial compression apparatus with special facilities for the measurement of the pore pressure; a schematic diagram of such an apparatus developed by Bishop (Bishop and Henkel, 1957) is shown in Figure 5.39. Its main features are as follows.

(a) Constant pressure is supplied to the cell from a self-compensating mercury control system (as shown).

(b) Pore water pressure in the sample is measured with a 'null indicator' pore pressure apparatus (as shown) or with electric resistance pressure transducers.

(c) In drained tests volume changes are measured with a simple burette (as shown) or with a special volume change indicator to which a constant back-pressure can be supplied from another mercury control system to ensure full saturation of the sample.

In order to obtain a clear picture of how the pore pressure responds to the different combinations of applied stress Skempton and Bishop have introduced

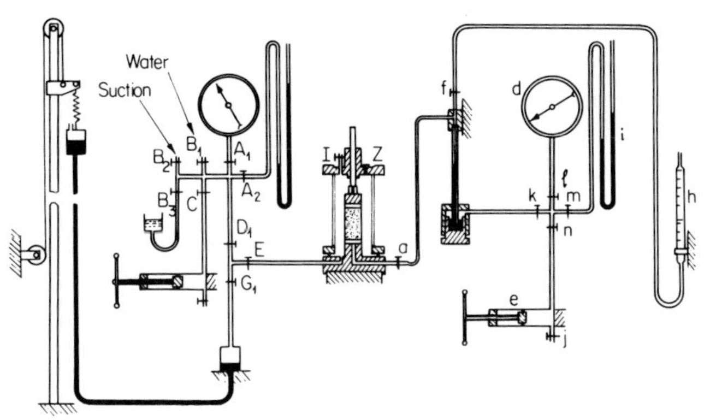

Figure 5.39. The layout of the apparatus for consolidated undrained tests on 38 mm diameter samples: with measurement of pore pressure (after Bishop and Henkel, 1957; by permission of Edward Arnold Ltd.)

the concept of *pore pressure parameters* (Skempton, 1954). This concept serves not only to explain the relationship between the different types of triaxial tests, but also provides a basis for estimating the magnitude of the pore pressures to be encountered in practical problems (see Section 5.2.7).

Let us consider a special case of an undrained cylindrical compression, i.e. the routine triaxial test. In this test the stress changes are usually made in two stages: (a) an increase in the cell pressure ($\Delta\sigma_3 = \Delta\sigma_2$), and (b) an increase in deviatoric stress $\Delta q = (\Delta\sigma_1 - \Delta\sigma_3)$. The resultant change in the pore pressure

Δu can be expressed in terms of the above stress changes with the help of two empirical pore pressure parameters A and B as follows:

$$\Delta u = B\{\Delta\sigma_3 + A(\Delta\sigma_1 - \Delta\sigma_3)\} \tag{5.48}$$

or for a more general case in which the stress changes are expressed in terms of an increase or decrease in the mean stress $\Delta\sigma_m = (\Delta\sigma_1 + 2\Delta\sigma_3)/3$ and deviatoric stress Δq:

$$\Delta u = B\{\Delta\sigma_m + (A - \tfrac{1}{3})|\Delta\sigma_1 - \Delta\sigma_3|\} \tag{5.49}$$

Note that the absolute value of $\Delta q = |\Delta\sigma_1 - \Delta\sigma_3|$ is taken.

In order to appreciate the physical significance of the two pore pressure parameters let us consider the two stages of an undrained triaxial test in more detail.

In the first stage of the test an all-around stress increase of the amount $\Delta\sigma_3$ is applied to the sample which induces an increase in the pore pressure of Δu_b. This rise in the pore pressure results in a compression (volume change) of the pore space equal to

$$\Delta V_p = m_p \Delta u_b$$

where m_p is the volumetric compressibility of the pore space due to an all-around increase in pressure. On the other hand, the increase in the effective stress which is equal to the difference $\Delta\sigma_3 - \Delta u_b$ causes a compression of the soil structure equal to

$$\Delta V_s = 3m_c(\Delta\sigma_3 - \Delta u_b)$$

where m_c is the uniaxial volumetric compressibility of the soil skeleton. If we assume the solid particles to be incompressible, then for the undrained condition ΔV_p must be equal ΔV_s and hence

$$\Delta u_b = \frac{1}{1 + m_p/3m_c}\Delta\sigma_3 = B\Delta\sigma_3 \tag{5.50}$$

where B is a coefficient dependent on the soil properties m_p and m_c.

For a saturated soil m_p is the compressibility of water which in comparison with m_c can be considered as very small; hence the coefficient B may be taken as equal to 1·0. For partly saturated soils the compressibility of the pore fluid (water and gases) is large and B is less than 1·0. The value of the coefficient B for partly saturated soils depends on the degree of saturation and is therefore stress dependent.

In the second stage of the test the major principal stress is increased by $\Delta q = \Delta\sigma_1 - \Delta\sigma_3$ and the pore pressure in the soil undergoes a corresponding change Δu_a. To evaluate the quantity Δu_a we shall consider the deformations of the sample under the stress increase.

The effective stress in the axial direction is increased by $\Delta q - \Delta u_a$ while the radial effective stresses are decreased by Δu_a. The increase in the axial stress produces a volume decrease equal to $m_c(\Delta q - \Delta u_a)$, whereas the reduction of the radial stresses produces an expansion of the soil equal to $2m_e(\Delta u_a)$, where m_e is the uniaxial volumetric expansibility of the soil. Let us consider now a saturated soil. In the undrained state the total volume change is equal to zero and we have

$$m_c(\Delta q - \Delta u_a) = 2m_e\Delta u_a$$

On rearrangement we obtain

$$\Delta u_a = \frac{1}{1 + 2m_e/m_c}\Delta q = A(\Delta\sigma_1 - \Delta\sigma_3) \tag{5.51}$$

where A is a coefficient dependent on the soil properties m_c and m_e.

For normally consolidated soils (such as soft clays) compressibility is large compared with the expansibility and A may approach 1·0, whereas for overconsolidated soils, such as stiff clays and dense sands, A is usually very small (Figure 5.40(d)). The value of A also depends on the proportion of the applied stress to the failure stress (Figure 5.40(c)); it is usually quoted for the failure condition and is then denoted by A_f.

It can be shown that for partly saturated soils the relationship between Δq and Δu_a is approximately given by the following expression:

$$\Delta u_a = AB\Delta q \tag{5.52}$$

By combining Equations (5.50) and (5.52) the general expression for the resultant pore pressure change (Equation (5.48)) is obtained.

Typical results of consolidated undrained tests (CU) on samples of normally consolidated and overconsolidated soils (S_r = 1·0 and B = 1·0) are shown in Figure 5.40: the point of failure is taken as occurring at either $(\sigma_1 - \sigma_3)_{max}$, or at $(\sigma'_1/\sigma'_3)_{max}$. Interpretation of the results of CU tests is shown in Figure 5.41.

The use of the Skempton–Bishop method is gaining popularity in and outside the British Isles, because of its relative simplicity and because of the great versatility of the equipment. With the introduction of pressure transducers and electronic internal load cells many of the difficulties and inaccuracies associated with pore pressure measurements and triaxial testing have been eliminated although the cost of the equipment, particularly of the recording instruments, has considerably increased.

5.4.6.4. Work of Borowicka and Skempton–Residual Strength

Borowicka (1961) has carried out a series of shear box tests on remoulded normally consolidated clays and undisturbed clays in which the samples were repeatedly sheared, by reversing the direction of shear (multiple shearing), and so were subjected to large shear displacements. The results have shown that the shear strength of certain soils decreased continuously until a very low limiting value was achieved (angles of internal frictional resistance as low as 6° were recorded). It was also observed that in most cases, after completion of the shear tests, the soil specimen could be broken into two halves with the surface of the shear plane exhibiting a shiny slick appearance, identical with the slickensides observed in

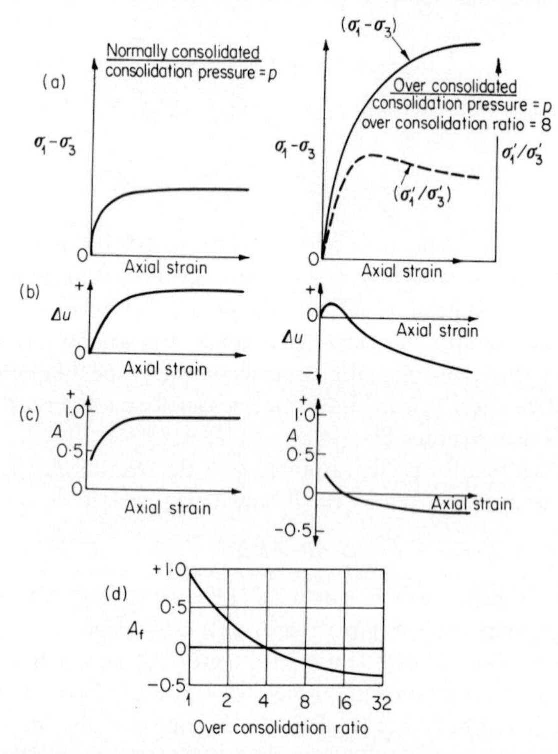

Figure 5.40. The change in pore pressure during the application of the deviator stress; typical results for normally and overconsolidated clay samples (after Bishop and Henkel, 1957; by permission of Edward Arnold Ltd.): (a) deviator stress; (b) pore pressure change; (c) value of parameter A, plotted against axial strain; (d) A_f, the value of A at failure, plotted against overconsolidation ratio.

landslide materials. Shear strength obtained from analysis of active or reactivated landslides were found to correlate reasonably well with the low values measured in the laboratory.

The results of work carried out in Britain (Skempton, 1964; Skempton and Hutchinson, 1969) are summarized in Figures 5.42 and 5.43.

In Figure 5.42 typical stress–strain curves are plotted for normally consolidated and overconsolidated clays tested under drained conditions. In both cases shear strength (under a given effective normal stress) reaches a peak value and

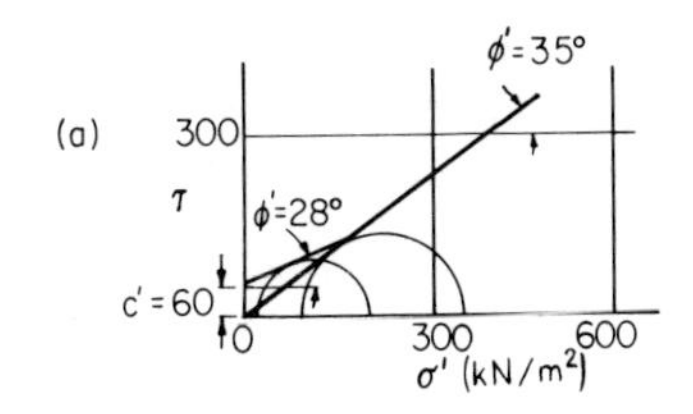

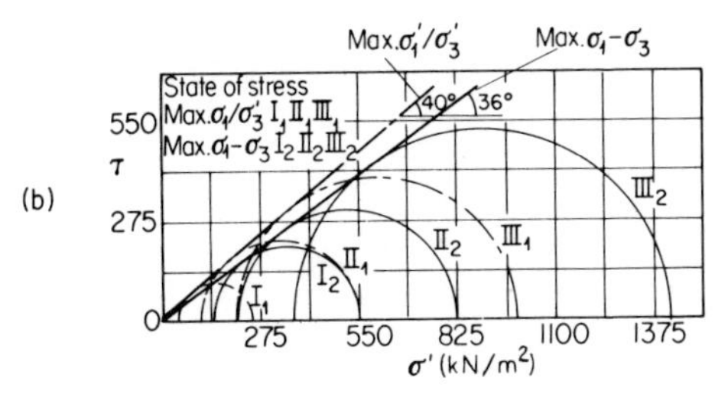

Figure 5.41. Failure envelopes obtained from CU tests: (a) Mohr circles in terms of effective stresses for the same soil in normally consolidated and overconsolidated state; (b) comparison of failure envelopes for partly saturated compacted moraine soil at $(\sigma_1'/\sigma_3')_{max}$ and at $(\sigma_1 - \sigma_3)_{max}$ from tests I, II, and III at different cell pressures (after Bishop and Henkel, 1957; by permission of Edward Arnold Ltd.).

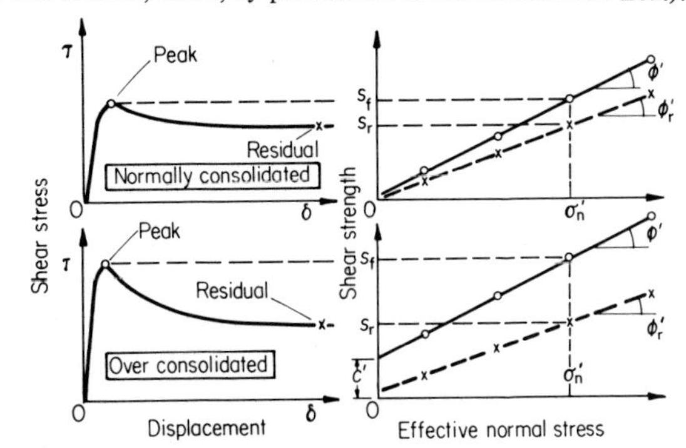

Figure 5.42. Simplified shear strength properties of clay (after Skempton, 1964; by permission of the Council of the I.C.E.).

then, as the displacements increase, decreases to a nearly constant value known as the residual strength.

Soft silty clays may show little difference between peak and residual but as the clay content increases the difference becomes more pronounced, even for normally consolidated soils; the decrease in strength is associated with reorientation of clay particles along the slip surface. Most stiff, i.e. overconsolidated, clays show a very pronounced decrease in strength from peak to residual which is partly due to particle reorientation effect and partly due to an increase in water content, resulting from dilatation of soil within the zone of shearing. These effects increase with clay content and the degree of overconsolidation.

The influence of clay mineral content on the magnitude of the residual angle of internal friction ϕ_r' is shown in Figure 5.43 in which ϕ_μ is the angle of friction between flat crystal faces of different minerals.

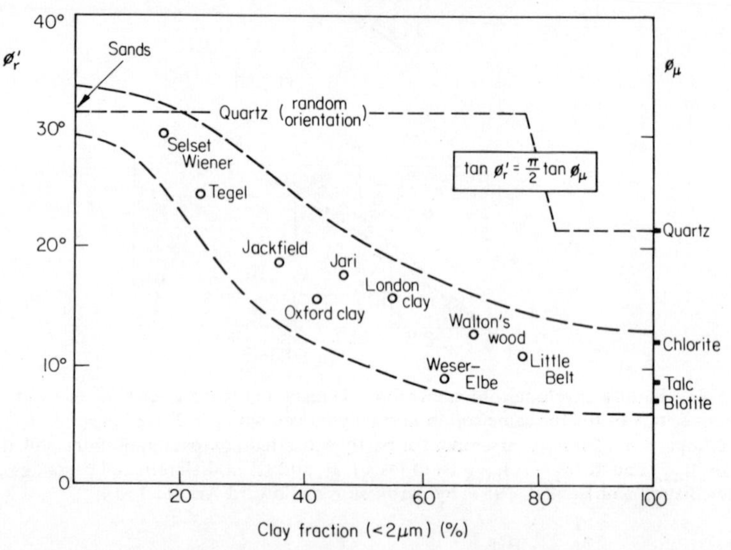

Figure 5.43. Decrease in ϕ_r' with increasing clay fraction (after Skempton, 1964; by permission of the Council of the I.C.E.).

5.4.6.5. Work of Kézdi–Time Effects

Kézdi (1962) discusses the influence of time on the deformation of soils subjected to shear loading and points out significant differences between the behaviour of cohesionless and cohesive soils under this type of loading.

In cohesionless soils there is a unique relationship between shear stresses and deformations, which cease almost immediately after application of loading. Under given σ_n and τ specific deformation is obtained or, conversely, a sand, given a certain shear deformation under a given σ_n, can only mobilize a specific shear resistance τ. This can be illustrated with the results of simple shear box tests on samples of sand having the same voids ratio but sheared under different normal stresses σ_n (Figure 5.44). As long as the shear strength is not exceeded ($\tau < \tau_f$), the sheared sand exhibits deformations which quickly attain their maximum value and cease; on reaching the shear strength ($\tau = \tau_f$), however, continuous deformations of the sand take place.

Deformation of sand during shearing, with relation to time, can be illustrated as shown in Figure 5.45.

The state of deformations in cohesive soils with respect to time and shear stress level is similar, but only for shear stresses below a certain threshold value τ_0; above that value the state of rest is not achieved after the initial deformations,

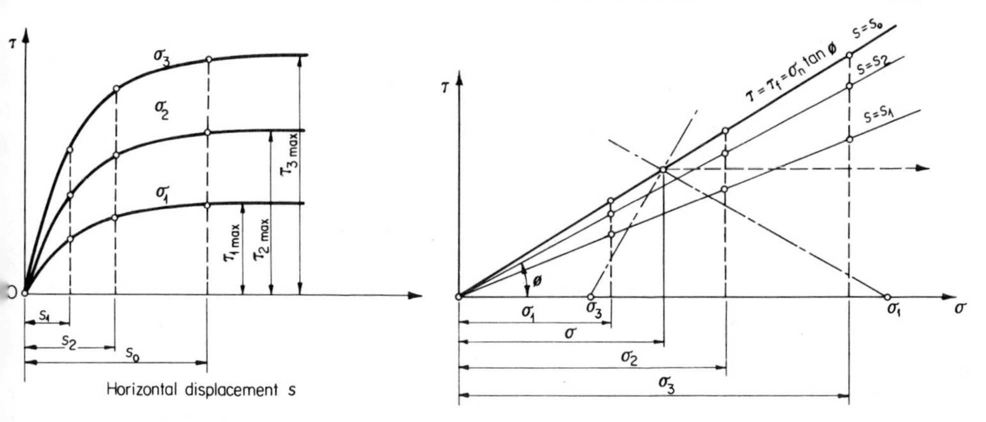

Figure 5.44. Typical stress–strain relationship obtained from a shear box test on sand.

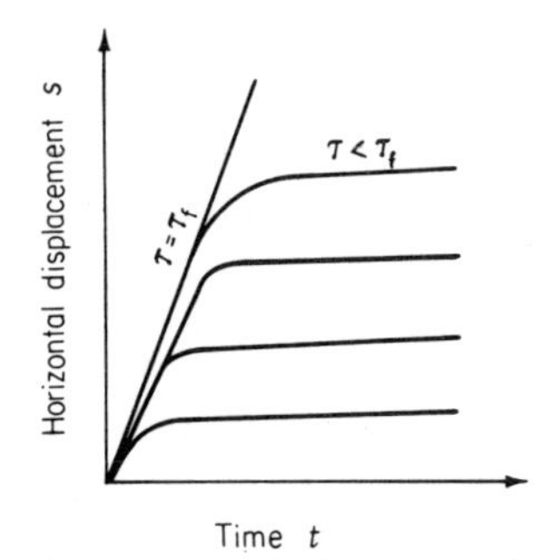

Figure 5.45. Deformation–time relationship for sand subjected to different levels of shear stress (normal stress constant).

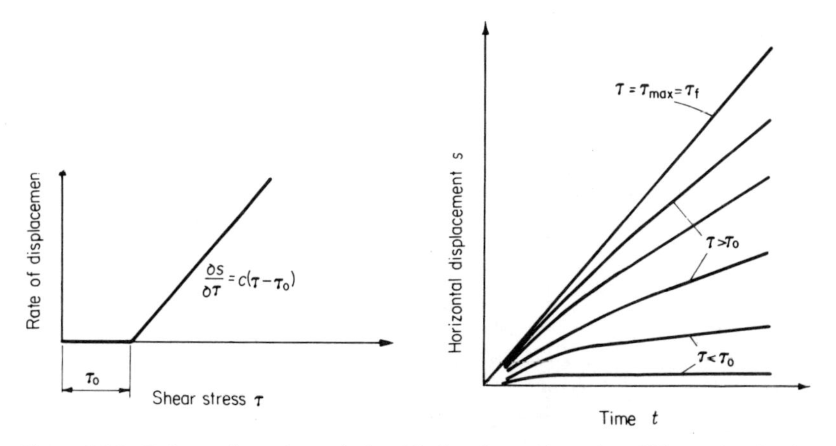

Figure 5.46. Deformation–time relationship for clay subjected to different levels of shear stress (normal stress constant).

but continuous movement (creep) occurs at a rate which is, approximately, directly proportioned to the difference $\Delta = \tau - \tau_0$.

For cohesive soils the equivalent of Figure 5.45 will therefore be Figure 5.46.

The threshold (or yield) shear stress τ_0 is sometimes referred to as the basic shear strength of the soil, because cohesive soils will only remain in the state of rest when shear stresses $\tau < \tau_0$; if shear stresses τ exceed τ_0, then creep will occur as can sometimes be observed in clay slopes or in structures founded on cohesive soils, e.g. continuous settlement of the famous Tower of Pisa.

The progressive shear failure of cohesive soils is also the result of creep deformations; high shear stresses at a point within a certain zone of a soil body induce creep deformations which gradually result in reduction of the shear strength of the soil to its residual value and lead to a local failure. Part of the load, initially carried by that zone, is now transferred to the neighbouring zones of the soil in which the level of the shear stresses is now increased beyond the threshold value and the whole process is now repeated until a slip surface forms, along which the low shear strength is no longer sufficient to maintain the overall equilibrium and a collapse takes place.

The parameter τ_0 can therefore be considered as one of the important properties of the soil and should be taken into consideration when dealing with stability problems.

It can be seen from the above review of the different methods of shearing and different cases of shear deformations of soils that each particular problem will require the selection of an appropriate method of testing which will closely correspond to the actual *in situ* stressing of the soil. It must also be strongly emphasized that any results of shear tests, in a form of strength parameters c and ϕ, should always contain a detailed description of the methods of testing and interpretation.

5.4.6.6. *Strength of Cohesive Soils in Unconfined Compression and Their Sensitivity*

As can be seen from Table 5.7 and, in particular, from undrained test results on saturated soft to firm silty clays or clays, the angle of internal shearing resistance ϕ_u for these materials is either very small or actually equal to zero;

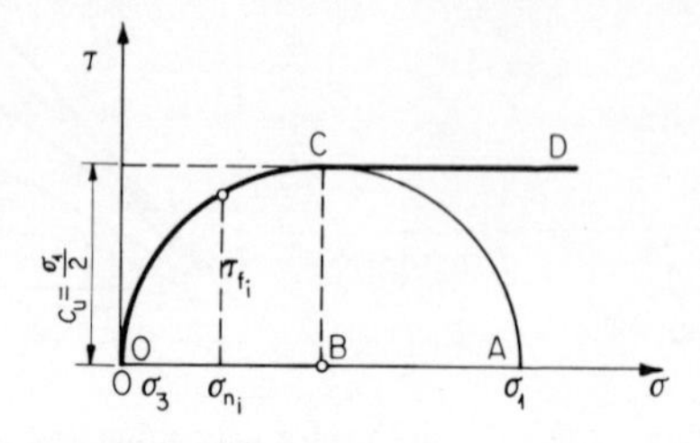

Figure 5.47. Determination of shear strength of clay from unconfined compression test.

therefore their shear strength is principally due to the apparent cohesion c_u. Assuming that $\phi_u = 0$, it is possible to determine the shear strength of a cohesive soil in one-dimensional (unconfined) compression test on an undisturbed sample of soil; the Mohr circle (Figure 5.47) is plotted by taking

$$\sigma_3 = 0 \text{ and } \sigma_1 = \frac{Qh'}{A_0 h_0}$$

where A_0 = initial cross-sectional area of sample
h_0 = initial height of sample
h' = final height of sample at failure
Q = axial load at failure

If $\phi_u = 0$, then

$$c_u = \frac{\sigma_1}{2} \tag{5.53}$$

However, in the stress range $\sigma_n < \overline{OB}$ the shear strength τ_f should be determined from consideration of a tangent to the arc OC.

If another sample of the same soil is tested in a remoulded state, i.e. if its natural structure has been destroyed but it has been reconstituted to the same natural density and water content, then the recorded strength may be considerably lower. The ratio of the compressive strength σ_1 of the undisturbed sample to the strength σ_{1R} of the remoulded one is known as the sensitivity of cohesive soils:

$$S_t = \frac{\sigma_1}{\sigma_{1R}} \tag{5.54}$$

According to Terzaghi clays can be classified as

insensitive, if $1 < S_t \leqslant 4$
sensitive, if $4 < S_t \leqslant 8$
very sensitive, if $S_t > 8$

The decrease in the strength of the remoulded soil is principally due to destruction of any interparticle bonds that existed in the natural soil and also due to destruction of the soil structure which may possibly affect the interparticle forces which control cohesive resistance.

On the basis of observed reduction in strength of certain silty soils due to disturbance of thixotropic bonds (see Section 2.4) the authors suggest that sensitivity of such soils to vibration and compaction should also be considered.

5.4.7. STATE OF LIMITING EQUILIBRIUM OF SOIL MEDIA

A state of limiting or plastic equilibrium exists in a soil when the following condition is satisfied:

$$|\tau| = \tau_f$$

This condition can be illustrated graphically with the help of the Coulomb strength envelope and the Mohr circles of stress (Figure 5.48).

Let us consider a simple triaxial compression test in which the soil sample is subjected to a constant cell pressure σ_3 while the vertical deviatoric stress q is increased until it reaches a certain maximum value q_{max} which cannot be exceeded but which, if maintained, will result in continuous deformation

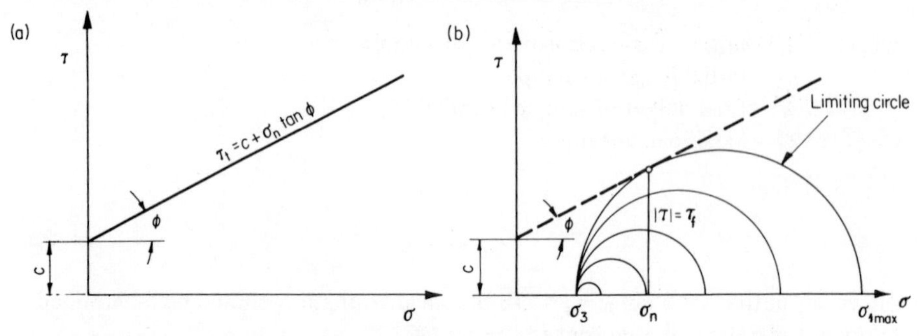

Figure. 5.48. Graphical illustration of the state of plastic equilibrium: (a) the Coulomb strength envelope; (b) the limiting Mohr circle of stress.

(yielding) of the sample. This implies that the sample has been sheared along a certain slip plane (or along a number of slip planes). As the deviatoric stress increases so does the $\sigma_1 = q + \sigma_3$ and hence a series of stress circles can be plotted with the limiting one defined by σ_3 and $\sigma_{1\,max}$ (Figure 5.48). Since this circle represents a state of stress which has induced limiting equilibrium within the sample, it must be tangential to the failure envelope because at the point of contact the necessary condition is satisfied:

$$|\tau| = \tau_f$$

It is possible to combine the geometrical properties of Mohr's stress circle with the equation of Coulomb's failure envelope, and to derive analytical relationships between σ_1, σ_3, τ_f, σ_n, and strength parameters c and ϕ. Such relationships were first derived by Mohr and are therefore often referred to as the Coulomb–Mohr failure criterion.

5.4.7.1. Limiting Equilibrium of Cohesionless Soils

In cohesionless soils $c = 0$ while $\phi > 0$ (Figure 5.49).

Considering triangle OAB (Figure 5.49(a)) it can be written that

$$\sin\phi = \overline{AB}/\overline{OA}$$

However, $\overline{AB}$ is the radius of the limiting circle and therefore $\overline{AB} = (\sigma_1 - \sigma_3)/2$ while $\overline{OA} = (\sigma_1 + \sigma_3)/2$; hence, substituting into the above equation

$$\sin\phi = (\sigma_1 - \sigma_3)/(\sigma_1 + \sigma_3) \qquad (5.55)$$

This is one form of the Coulomb–Mohr failure criterion for cohesionless soils.

The normal stress σ_1 (the major principal stress) can be expressed in terms of σ_3 and ϕ with the help of Figure 5.49(b).

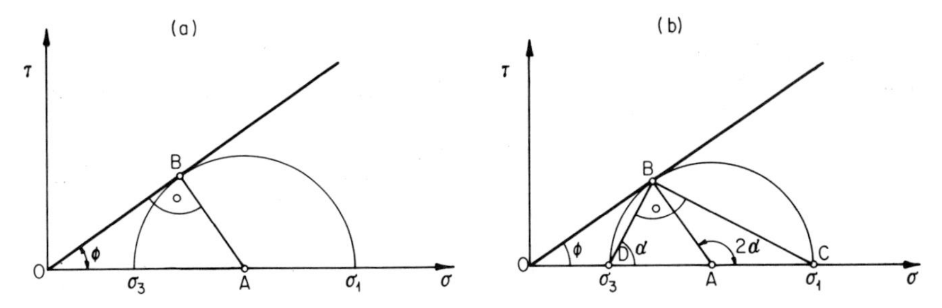

Figure 5.49. Limiting equilibrium in cohesionless soils: (a) diagram i; (b) diagram ii.

From similar triangles OBC and ODB

$$\overline{OC}/\overline{OB} = \overline{OB}/\overline{OD} = \overline{BC}/\overline{DB}$$

and considering that $\overline{BC}/\overline{DB} = \tan\alpha$ it can be written that

$$\overline{OC}/\overline{OD} = \tan^2\alpha$$

Taking now into consideration that $\overline{OC} = \sigma_1$ and $\overline{OD} = \sigma_3$ and substituting into the above

$$\sigma_1 = \sigma_3 \tan^2\alpha$$

From triangle OAB, $2\alpha = 90^\circ + \phi$ or $\alpha = 45 + \phi/2$, therefore on substitution

$$\left.\begin{aligned} \sigma_1 &= \sigma_3 \tan^2\left(45 + \frac{\phi}{2}\right) \\ &\text{and} \\ \sigma_3 &= \sigma_1 \tan^2\left(45 - \frac{\phi}{2}\right) \end{aligned}\right\} \quad (5.56)$$

If the state of limiting equilibrium is defined by stress components referred to some orthogonal axes x and z, then the principal stresses σ_1 and σ_3 can be expressed in terms of these components with the help of Figure 5.50.

It can be seen that

$$\sigma_1 = \overline{OA} + \overline{AC} \quad \text{and} \quad \sigma_3 = \overline{OA} - \overline{AC}$$

From triangle ACD,

$$\overline{AC} = \sqrt{\left\{\left(\frac{\sigma_z - \sigma_x}{2}\right)^2 + \tau_{xz}^2\right\}}$$

which when substituted into the above equations gives

$$\sigma_{1,3} = \frac{\sigma_z + \sigma_x}{2} \pm \sqrt{\left\{\left(\frac{\sigma_z - \sigma_x}{z}\right)^2 + \tau_{xz}^2\right\}}$$

Substituting these values of σ_1 and σ_3 into Equation (5.55) a general expression of the Coulomb–Mohr failure criterion for cohesionless soils is obtained:

$$\frac{\sqrt{\{(\sigma_z - \sigma_x)^2 + 4\tau_{xz}^2\}}}{\sigma_z + \sigma_x} = \sin\phi \qquad (5.57)$$

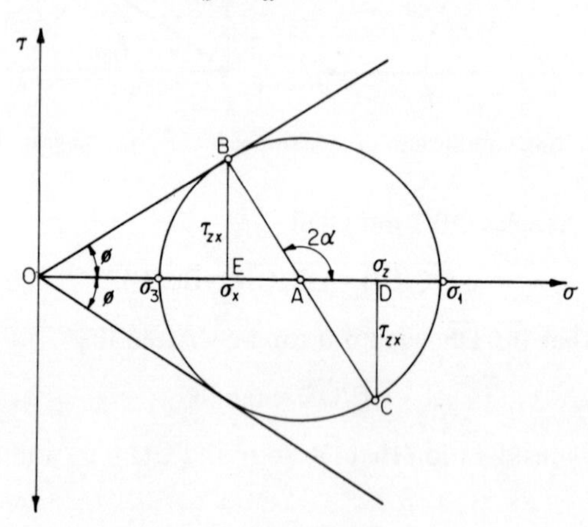

Figure 5.50. Limiting equilibrium in cohesionless soils: diagram iii.

5.4.7.2. Limiting Equilibrium of Cohesive Soils

For cohesive soils both c and $\phi > 0$.

In order to modify Equations (5.55), (5.56), and (5.57) for the use with cohesive soils it is necessary to substitute into them $\sigma_1 + c\cot\phi$ and $\sigma_3 + c\cot\phi$ for σ_1 and σ_3 respectively (Figure 5.51). By introducing $\sigma_c = c\cot\phi$ the following expressions for the Coulomb–Mohr failure criterion for cohesive soils are obtained:

$$\sin\phi = (\sigma_1 - \sigma_3)/(\sigma_1 + \sigma_3 + 2\sigma_c) \qquad (5.58)$$

$$\sigma_1 = \sigma_3 \tan^2\left(45 + \frac{\phi}{2}\right) + 2c\tan\left(45 + \frac{\phi}{2}\right)$$

and

$$\sigma_3 = \sigma_1 \tan^2\left(45 - \frac{\phi}{2}\right) - 2c\tan\left(45 - \frac{\phi}{2}\right) \qquad (5.59)$$

$$\left.\begin{aligned}&\frac{\sqrt{\{(\sigma_z-\sigma_x)^2+4\tau_{xz}^2\}}}{\sigma_x+\sigma_z+2\sigma_c}=\sin\phi\\ &\text{or}\\ &(\sigma_z-\sigma_x)^2+4\tau_{xz}^2=(\sigma_x+\sigma_z+2c\cot\phi)^2\sin^2\phi\end{aligned}\right\}\quad(5.60)$$

Also from Figure 5.51 the following relationship can be obtained:

$$\overline{BC}=\overline{AB}-\overline{AC}$$

or

$$\frac{\sigma_1+\sigma_3}{2}\sin\phi=\frac{\sigma_1-\sigma_3}{2}-c\cos\phi \quad (5.61)$$

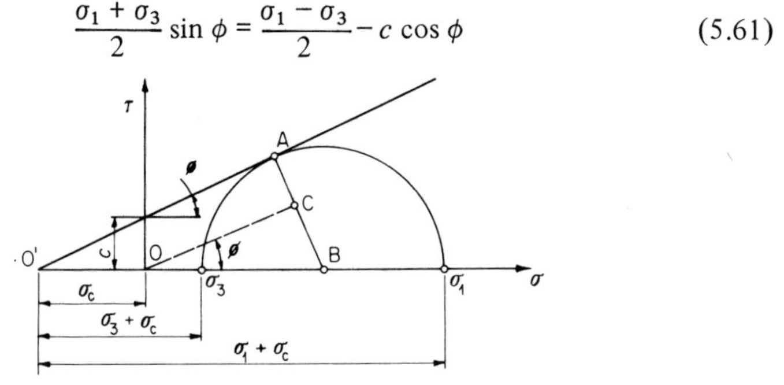

Figure 5.51. Limiting equilibrium in cohesive soils: diagram iv.

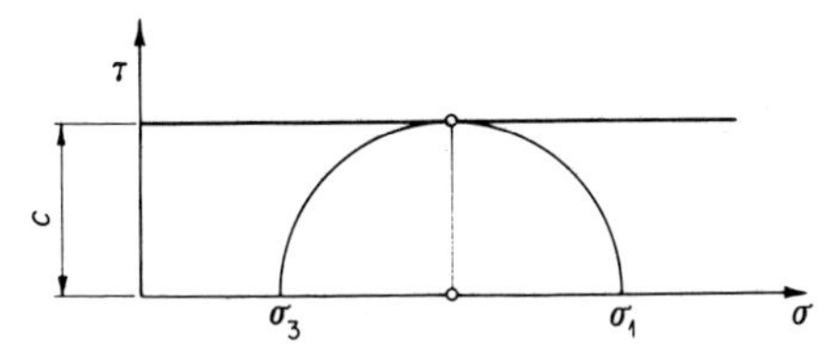

Figure 5.52. Limiting equilibrium in ϕ = 0 soils: diagram v.

For cohesive soils with $c > 0$ but $\phi = 0$ the Coulomb failure envelope becomes horizontal (Figure 5.52) and the above equations reduced to

$$\left.\begin{aligned}&\sigma_1=\sigma_3+2c\\ &\text{and}\\ &\sigma_3=\sigma_1-2c\end{aligned}\right\}\quad(5.62)$$

5.4.7.3. Direction of Planes of Failure

If at a given point within the soil mass shear failure occurs, then the stresses acting on a certain pair of planes through that point have reached limiting equilibrium. The directions of these planes (known as the planes of failure) can easily be established from the geometry of the Mohr circle of stress. Since we are dealing with a case of axial symmetry one angular dimension with respect to the line of action of one of the principal stresses is sufficient to define such a plane, and therefore angles α_1 and α_2 will be used to define the pair of failure planes with respect to the direction of σ_3.

The stresses on the planes of failure must satisfy the Coulomb–Mohr failure criterion expressed, depending on type of soil under consideration, by one of Equations (5.56) to (5.62). As can be seen in Figure 5.53 the only two points that satisfy these conditions are the points of contact between the Coulomb failure envelope and the Mohr circle of stress. The planes of action of normal

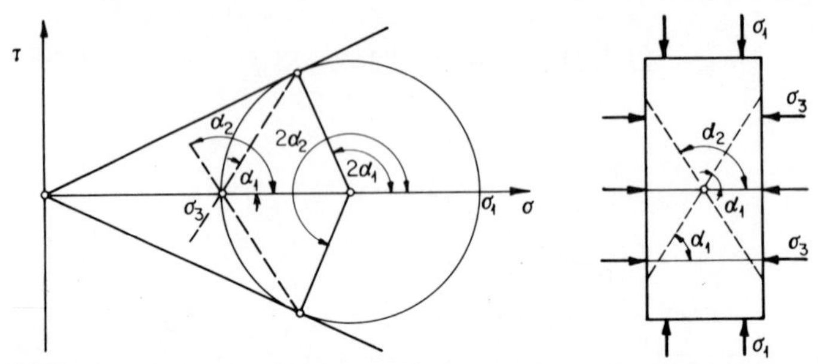

Figure 5.53. Direction of planes of failure in simple triaxial compression test.

stresses represented by these points are inclined at $\alpha_1 = 45° + \phi/2$ and $\alpha_2 = 135° - \phi/2$ to the line of action of σ_3.

Straight lines drawn through the given point in soil at inclinations α_1 and α_2 to the direction of the minor principle stress give the directions of the planes of failure, also known as slip planes.

The stress ratio σ_1/σ_3 reaches its maximum possible value at all points within these planes. The two planes intersect at an angle $\omega = 90° - \phi$ which is only dependent on the angle of internal shearing resistance of a given soil.

5.5. Additional Remarks

Mechanical properties of soil should be determined from tests on undisturbed samples, the exceptions being the determination of sensitivity of clays and tests on fill materials in which remoulded samples are used. It must, however, be pointed out that extraction of good-quality undisturbed samples and preservation of their natural state prior to testing is very difficult.

It was observed, for example, that on prolonged storage of samples of cohesive soils stored in steel tube samplers and sealed with paraffin wax water content has decreased and has resulted in values of c and ϕ being higher than in the natural state. Mechanical disturbance of soil structure or swelling due to absorption of water will on the other hand result in lower values of c, ϕ, and E, the stiffness modulus being particularly sensitive to this type of disturbance. Owing to their layered structure varved clays are particularly prone to swelling; the thin layers of clay sandwiched between saturated layers of silt or fine sand absorb water from them and may change in consistency from very stiff to firm or even soft, rendering the testing of such materials futile.

Silty sands and silts can be loosened by seepage of ground water into the borehole prior even to extraction of the sample. Loose sands can be densified during sampling as well as during transportation. Freezing of cohesive soils will result in reduction of cohesive and internal shearing resistance.

The best solution to the problem is undoubtedly the *in situ* testing of soils, e.g. loading tests, shear vane test, etc. (see Section 6.3). Testing of soils in field laboratories, immediately on extraction of samples, is also strongly recommended.

It is necessary to emphasize that correct results from laboratory tests can be obtained, in the case of determination of compressibility (oedometer test) only for samples of soft and firm soils, while in the case of shearing in triaxial apparatus only for samples of fairly strong soils (stiff or very stiff consistency). It is suggested that in all other cases laboratory testing should be replaced with *in situ* testing (see Section 6.3).

Particular difficulties are encountered in testing of samples of laminated soils or of soils with pre-existing slip planes (slickensides). It is important to ensure that in such cases the direction of the induced plane of failure coincides with the pre-existing plane of weakness.

6

Site Investigations

6.1. General Considerations

6.1.1. OBJECT AND PHASES OF SITE INVESTIGATIONS

Site investigations* are usually carried out prior to the design of roads, buildings, or other civil engineering works in order to establish the soil and ground water conditions on a given site; 'spot-check' type of site investigations should be carried out again, during the execution of earthworks, to check whether the soil and ground water conditions correspond to those assumed in the design.

Site investigations prior to the design of major works are usually carried out in three phases.

(1) Preliminary investigations–to assess the suitability of a given site for the proposed works or to select the most suitable site or route from several alternative sites (routes).

(2) General investigations–to select areas on a given site of the most favourable ground conditions.

(3) Detailed investigations–to obtain technical data necessary for the design and construction of the works (after approval of the location of buildings or choice of the route).

In certain justified cases of minor works some phases of the site investigations may be omitted; for example, if general investigations show that the geological structure of a given site is uniform and the ground water table is at a considerable depth below the surface, then the detailed investigation may not be necessary.

Preliminary investigations basically involve the study of the geology of a given area on the basis of available geological publications and actual inspection of the site or the route; in certain cases exploratory borings and simple subsurface soundings are carried out in important locations (deep cuttings, landslide areas,

* This subject is discussed very thoroughly in the British Code of Practice CP 2001 (1957).

river crossings, etc.). Particularly useful at this stage are the geological maps and handbooks on the regional geology of Great Britain compiled by the Geological Survey of Great Britain; most parts of the British Isles are covered by 1 : 63360 (1 in to 1 mile) scale maps (solid and drift edition) and some of the more developed areas by 1 : 10560 (6 in to 1 mile) scale maps. Other sources of information may be found in Dumbleton and West (1971). In mining areas the records of the National Coal Board should be consulted for information on abandoned, active, and prospective workings.

General investigations involve the preparation of a geotechnical map of the area under consideration on the basis of published information (geological maps, sheets, memoirs, etc.), inspection and mapping of outcrops, results of supplementary investigations (aerial photographs, geophysical and biological surveys), and exploratory borings on 100–300 m grid in the case of a building site and at 300–500 m centres along the route of a road; the depth of borings is established with reference to the type of the proposed works and anticipated soil conditions. Where possible, simple subsurface soundings are made prior to borings so that the weakest strata are located and can then be investigated thoroughly during the boring.

All the collected information is plotted on the geotechnical map and areas of similar ground conditions are indicated.

Detailed investigations should accurately establish the geological structure of the investigated area and should give complete details of the ground and ground water conditions necessary for the design and construction of the works (structural foundations, roads, hydroelectric projects, etc.).

In order to ensure that the results of laboratory and *in situ* tests refer to the soil strata which are the most significant in determination of the allowable bearing stresses for a given structure the authors recommend that at each borehole location subsurface soundings of the type described in Section 6.3 should be made prior to boring.

Stand pipes or piezometers should be installed in boreholes for the purpose of observation of the fluctuations of the ground water table, particularly during the construction period, to enable the selection of the best method of dewatering (if necessary).

For foundation engineering purposes the geotechnical sections should indicate, in addition to the geologically different strata, any geotechnical variations which may be present within any stratum. For road construction purposes typical sections of similar soil and ground water conditions should be indicated. A summary of the *in situ* and laboratory tests results should give complete quantitative and qualitative information about the soils within the individual layers indicated on the sections. Ground water tables (normal and the highest) should be indicated on the sections with a continuous line.

A site investigation report should contain, in addition to the location plan, geotechnical sections, and summary of test results, a description of the site (topographical, morphological, geological, and hydrological) and of the type and condition of existing buildings and roads. It should discuss, also, the probability of landslides, the movements of the ground water table, and the possibility of utilization of the local materials for construction purposes.

The object and scope of the site investigation work is therefore very broad. The co-operation of geologists is indispensable, because the preliminary and general phases of a site investigation basically consist of a geological survey and in the detailed phase their advice is required in preparation of the geotechnical sections.

6.1.2. SPACING AND DEPTH OF BOREHOLES

The purpose of the exploratory investigations is to obtain accurate information about the actual ground and ground water conditions at the site of a new structure or along the route of a new road. The information should be sufficient to enable one to prepare characteristic geological sections of the ground beneath the entire proposed structure or to prepare a three-dimensional model of the strata and of the ground water levels.

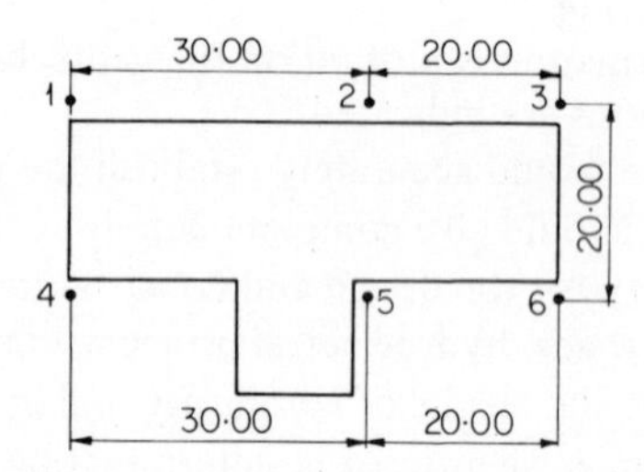

Figure 6.1. Spacing of borings for a building.

To achieve this the spacing of borings should not be based only on the type of structure but also on the uniformity of the ground conditions. Usually a preliminary estimate of the spacing is made which is subsequently either decreased, if additional data are necessary, or increased if the soil conditions are found to be approximately the same in all the borings. Spacing should be smaller in areas of high concentration of heavy loads and greater in less critical areas.

For individual buildings of plan area smaller than 300 m^2 at least three borings should be made (not in one line), for larger buildings at least five borings (one at each corner and one in the middle) but not more than 30 m apart in uniform ground conditions (Figure 6.1) and closer in irregular conditions. For preliminary estimates the number of borings can be taken according to Table 6.1.

In investigations for roads or large industrial or housing estates the number of boreholes should be limited, for economic reasons, to the essential minimum. The locations of boreholes should be chosen with reference to geological maps and topographical features which, to a certain degree, reflect the changes in the nature of soils beneath the ground surface; geophysical seismic and resistivity methods can be used to bridge the gaps between borings and to pin-point unexpected irregularities.

Table 6.1. Number and spacing of borings for site investigation work

Phase of investigation	Geological structure	Number and spacing of borings	Location of borings in the field
preliminary investigations (to assess suitability of a given site)	uniform	5 to 10 borings per km^2	depending on topography of the site—at highest, lowest, and intermediate points
	irregular or unknown	10 to 30 borings per km^2	
general investigations of a given site (for selection of areas with similar and most favourable ground and ground water conditions)	uniform	300 x 300 m	regular square network of borings situated parallel to contour lines
	irregular or unknown	100 x 100 m	
detailed investigations (for individual buildings whose locations have been fixed)	uniform	at least 3 borings 30 to 50 m apart	as regular as possible network of borings, modified to suit individual buildings and taking into consideration borings executed during preliminary or general investigations
	irregular or unknown	3 to 4 borings for each individual building 10 to 30 m apart	

In proposed cuttings soil conditions should be investigated in at least three locations along the centre line of the road, one or two of which may correspond to the points of transition between cutting and embankment. Along the sections of proposed embankments investigations are carried out at 200 m spacing, and, for short embankments, only at the lowest point.

When weak soils such as organic muds or peats are encountered, the spacing of borings should be decreased so that the depth of the soft soils can be accurately established along the centre line of the road and also along the boundaries of the proposed embankment. Under such circumstances the use of the vane/cone penetrometer is strongly recommended.

In areas which are doubtful with regards to the uniformity of soil conditions additional small-diameter shallow (3–5 m deep) borings can be made (between the main borings) to establish the extent of different strata.

The depth of borings, as in the case of spacing, does not depend only on the size and on the type of structure but also on the uniformity of the ground conditions.

For foundations of compact buildings on irregular soils borings should be taken down to depths at which the additional stresses induced by the foundations do not exceed 20% of the effective overburden stresses (Table 6.2); in the case of regular soil conditions or where competent strata are encountered at shallower depths, the borings need not be as deep as indicated in the table. When from the results of preliminary investigations (or from the local knowledge of soil conditions) it is anticipated that piled foundations are required, then the borings should penetrate between 5 to 7 m below the bottom horizon of the weak deposits.

Table 6.2. Depth of borings for foundations on irregular soils

Ratio of width to length of foundation $B:L$	Additional loading (above initial overburden stress) (kN/m^2)	Width of foundation (m)									
		1	2	3	4	5	6	8	10	15	20
		Depth of borings below foundation level (m)									
1 : 1	40	5·0	5·0	5·0	5·0	5·0	5·0	6·0	7·0	8·0	8·0
	60	5·0	5·0	5·0	5·0	5·0	6·0	7·0	7·0	9·0	10·0
	80	5·0	5·0	5·0	5·0	6·0	6·0	7·0	8·0	10·0	11·0
	100	5·0	5·0	5·0	6·0	6·0	7·0	8·0	9·0	11·0	13·0
	200	5·0	5·0	6·0	7·0	8·0	9·0	10·0	12·0	15·0	18·0
	300	5·0	5·0	7·0	8·0	9·0	10·0	12·0	14·0	18·0	21·0
1 : 2 to 1 : 3	40	5·0	5·0	5·0	5·0	5·0	6·0	6·0	7·0	8·0	9·0
	60	5·0	5·0	5·0	6·0	6·0	7·0	8·0	9·0	10·0	11·0
	80	5·0	5·0	5·0	6·0	7·0	8·0	9·0	10·0	12·0	13·0
	100	5·0	5·0	6·0	7·0	8·0	9·0	10·0	11·0	13·0	15·0
	200	5·0	6·0	8·0	9·0	10·0	11·0	13·0	15·0	18·0	22·0
	300	5·0	7·0	10·0	11·0	13·0	14·0	16·0	18·0	22·0	27·0
1 : 4 to 1 : ∞	40	5·0	5·0	5·0	5·0	6·0	6·0	7·0	7·0	8·0	9·0
	60	5·0	5·0	5·0	6·0	7·0	7·0	8·0	9·0	10·0	11·0
	80	5·0	5·0	6·0	7·0	7·0	8·0	9·0	10·0	12·0	13·0
	100	5·0	6·0	7·0	8·0	8·0	9·0	10·0	11·0	14·0	15·0
	200	6·0	8·0	9·0	11·0	12·0	13·0	15·0	17·0	20·0	23·0
	300	7·0	9·0	11·0	13·0	15·0	16·0	18·0	20·0	25·0	30·0

The co-ordinates and ground levels of exploratory borings and trial pits should be accurately established with reference to the Ordance Survey Grid and Datum Levels or in the case of minor investigations with reference to some permanent feature (e.g. kerb line) on the site; this information will enable the location of borings to be shown on the layout plan and geological sections.

6.2. Methods of Soil Exploration

Although the large variety of the present-day exploratory techniques differ considerably in detail, some of them having been developed specifically for soil investigation and some having been adopted from other fields of exploration (e.g. for mineral ores or oil), they can conveniently be grouped into three categories: (a) subsurface soundings which are designed to measure the variation in properties of soils or rocks *in situ* without any means of retrieving samples for visual inspection (e.g. static penetration test or seismic refraction and reflection methods); (b) exploratory borings which enable one to extract continuous or discrete samples of the investigated soils or rocks for visual and physical determination of their properties; (c) combination of exploratory borings with subsurface soundings to enable one to extract samples of the investigated soils for inspection and laboratory testing and also to measure their properties (or variation in their properties) *in situ*. Since in soil exploration one has to deal with a very wide range of site conditions most of the modern exploratory techniques fall into the third category and include a combination of subsurface soundings without borings (e.g. vane/cone penetration test), exploratory borings with soil sampling and *in situ* testing (e.g. standard penetration, plate-loading, or pressure meter tests), and exploratory excavations (trial pits) with soil sampling and *in situ* testing (e.g. plate-loading tests).

6.2.1. EXPLORATORY EXCAVATIONS AND BORINGS

Exploratory excavations (*trial pits*) are generally used for exploration to depths smaller than 3·5 m, but with the use of mechanical excavators, and in suitable ground

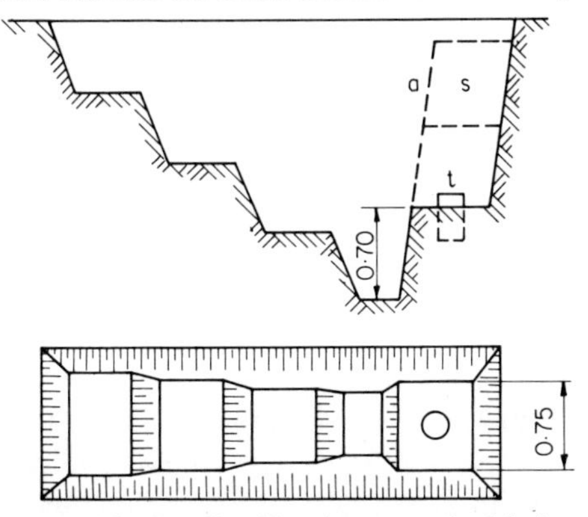

Figure 6.2. Details of hand-excavated trial pits.

conditions, the depth range can economically be extended down to approximately 8 m. Hand-excavated trial pits are usually about 2 m long at the top and narrow down to 0·6–0·8 m at the bottom (Figure 6.2).

The size and shape of mechanically excavated pits depends to a large extent on the type of excavator used; the use of timbering in trial pits is generally avoided. One of the narrow faces of the excavation (face 'a' in Figure 6.2) is kept almost vertical and, to ensure good natural lighting, is usually positioned to face south; undisturbed block samples 's' or tube samples 't' are taken from this face as shown in Figure 6.2.

To avoid weakening of the ground, trial excavations are made outside the outlines of the proposed foundation. Trial pits enable one to obtain a clear picture of the actual ground conditions and are indispensable in investigation of weathered rocks in which it is necessary to establish the degree and extent of weathering (spacing and size of joints) and the direction and magnitude of the dip. The use of trial pits is also recommended in investigations of fill materials and landslide areas where it may be necessary to locate narrow slip planes.

Exploratory borings (*boreholes*) are generally used for soil explorations to depths greater than 3·0 m, i.e. outside the usual depth range of the trial pits, or when difficult ground water conditions are encountered. The size (diameter) of boreholes generally varies between 50 and 250 mm and depends on the type of the investigated soil, the size of required samples, and the type of equipment used. In firm soils the hole remains open by arching but in soft clays and in sands below the ground water table it is kept open by inserting steel tubing (casing) or by keeping the hole filled with a drilling mud (a viscous suspension of bentonitic clay in water); drilling is done by augering, shelling, wash boring, or high-speed rotary drilling.

In the case of shallow holes augering can be done with a simple post-hole auger which consists of two curved blades that retain the soil as it is cut; for larger and deeper holes power augers with helical cutters can be used. In very soft soils or in gravels below the water table the shelling technique can be used for advancing the hole: a tube with a cutter and a non-return valve fitted at the bottom (known as the shell) is dropped to the bottom of a cased borehole and any material trapped inside it is brought to the surface. In clays, sands, and even soft rocks drilling can be done using the wash boring technique which is a combination of jetting and chopping with a chisel bit attached to hollow drill rods through which water is pumped; the cuttings dislodged by chopping are carried to the surface by the returning water. In overconsolidated clays and rocks, high-speed rotary drilling is used either for open-holing or for cutting cores; circulating water or drilling mud is used to remove the cuttings in the same way as in wash boring. Whichever of the above methods of drilling is used it must not be forgotten that the primary object of soil exploration is not to form holes in the ground but to secure samples of the encountered soils for identification and testing purposes; on the basis of these results characteristic properties of individual layers are determined and a complete picture is built up of the ground conditions on a given site.

Boreholes should be positioned close to the proposed foundations but outside their outlines. On completion of investigation they should be backfilled with medium cohesive soils, which should be well enough compacted to prevent ground water from entering foundation excavations or other dry strata.

The position of ground water levels in each water-bearing stratum (gravels in particular) should be measured and checked after 24 hours; the results should be recorded stating the dates of the two readings. In certain cases (particularly in locations close to rivers) the study of the ground water levels should be made over a period of several months or preferably over one complete year. The highest ground water levels can sometimes be located in boreholes by observing gley horizons, which can be recognized by their characteristic dark green or blue colour, or from information obtained from local inhabitants, e.g. details of variations of water levels in wells.

Useful information can also be obtained from local inhabitants with regards to the existence of springs and water-bearing strata on hillsides. This type of information supplements the data obtained from boreholes and trial pits.

Description of soils during explorations. During drilling of boreholes or excavation of trial pits samples of the encountered soils are examined and described (see Chapter 3); the details are entered in the drilling record. Other entries include the depths at which samples were taken, resistance to drilling, details and results of *in situ* tests, observations of ground water levels and depths of casing at the time of sampling, *in situ* testing, and ground water observations. The information contained in drilling records forms the basis for preparation of boreholes records and geological sections.

6.2.2. METHODS OF SOIL SAMPLING AND SAMPLE DISTURBANCE

During exploratory borings or trial pit excavations soil samples are taken for detailed description (macroscopic analysis) of the soils and for determination of their physical and mechanical properties. For description purposes disturbed soil samples are taken from each encountered stratum but at not more than 1·0 to 1·5 m intervals; these samples are usually placed in plastic bags or in wooden boxes in which they can readily be inspected. Representative samples from each stratum of cohesive soils are placed in air-tight jars (to preserve water content). For the purpose of determination of mechanical properties undisturbed tube samples are usually taken from those strata and depths which are relevant to the design of foundations (to a depth of one-and-a-half times the width of the foundation) and from deeper layers if their strength is low. The results of subsurface soundings with the vane/cone penetrometer, carried out prior to sinking of boreholes, enable one to select the most relevant depths at which undisturbed samples are to be taken or *in situ* tests are to be carried out; this procedure ensures that relevant information is available about all the encountered soils and frequently leads to considerable economy in undisturbed sampling and *in situ* and laboratory testing.

In the case of boring below the ground water level, loosening can take place of the soil below the advancing borehole if the level of the water inside the casing is lower than that of the ground water outside it. Particularly prone to this type of disturbance, or even liquefaction, are cohesionless or slightly cohesive soils; the soil to a depth of approximately two diameters of the casing is affected. In order to prevent this type of disturbance the water level inside the casing should be maintained several metres above the *in situ* ground water level. In such cases, if possible, undisturbed tube samples should be taken.

All samples should be labelled immediately after being taken from the borehole or trial pit. The label should indicate location of the site (or job number), boring number, sample number, depth from which sample was taken (e.g. from 1·5 to 1·8 m below GL), and the type of sample (disturbed or undisturbed).

The original consistency of cohesive soils can be estimated from the appearance of their samples stored in boxes: lumps of soils of hard or very hard consistency usually retain fairly sharp edges; the softer and more plastic the soils the more uniformly they fill the boxes, with individual lumps losing their identity, and on drying out a gap several millimetres wide appears along the walls of boxes (shrinkage). Alluvial muds and peats exhibit very considerable shrinkage and the gaps can be of the order of tens of millimetres.

It is not possible to take a completely undisturbed sample of any soil because the removal of the sample from the ground involves a change in the state of stress. Generally, however, provided the water content is not altered and the natural structure of the soil is not disturbed, the samples can be considered as undisturbed and will yield sufficiently accurate results for most practical purposes; changes in water content are prevented by careful handling and sealing of the samples (with paraffin wax) and by protection against extreme temperature changes; the structural disturbance of the soil can be reduced to a minimum by the use of correctly designed samplers and proper sampling techniques.

The main features of a well-designed thin-walled tube sampler are (a) low outside wall friction achieved by provision of outside clearance (Figure 6.4(a)), (b) low inside wall friction achieved by provision of inside clearance,* smooth finish to the sample tube and use of oil, (c) small ratio of the cross-sectional area of the tube or the cutter to that of the sample tube (area ratio A_r); the larger the area ratio is, the greater is the probability of disturbance because of the greater relative volume of the soil displaced by the sampler. Another feature, which plays an important part in minimizing the disturbance and at the same time helps to retain the sample in the tube during withdrawal, is a non-return valve provided in the drive-head; it allows free exit of air and water during driving and helps to create suction during withdrawal.

For structurally undisturbed samples, extracted under ideal conditions, the ratio of the sample recovery L to the depth of penetration of the sampler H should be equal to 1·0 at all stages of penetration. For most tube samplers,

* In the case of overconsolidated soils the provision of internal clearance may lead to greater disturbance (Ward *et al.*, 1959).

however, the recovery ratio varies as shown in Figure 6.3(b) (Hvorslev, 1949a). Initially, owing to the unavoidable disturbance of the soil within the immediate vicinity of the bottom of a borehole and owing to the existence of overburden stresses in the soil, there is a tendency for the soil to be displaced inwards and, therefore, during the initial stages $\Delta L/\Delta H$ (specific recovery) is greater than 1·0.

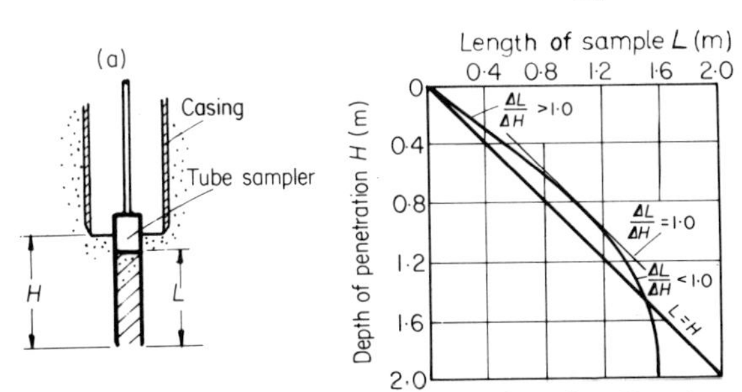

Figure 6.3. Structural deformation of soil samples (after Hvorslev, 1949a).

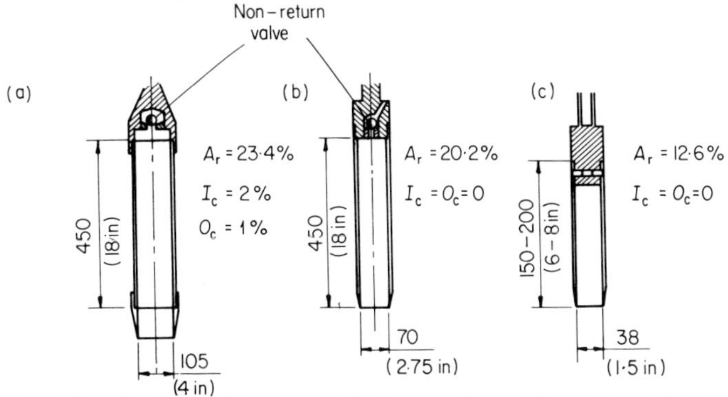

Figure 6.4. Details of simple thin-walled tube samplers: (a) used with shell and auger boring; (b) used with rotary core drilling; (c) usually used for subsampling for laboratory testing.

As the length H of the sample in the sampler increases, the frictional forces take over and begin to compress the soil so that $\Delta L/\Delta H$ drops below 1·0 and eventually reaches zero; at this stage soil ceases to enter the tube and the sampler with the enclosed soil behaves like a solid penetrometer.

Apart from the design details of a sampler the method of forcing the sampler into the ground also has a considerable effect on the sample disturbance; samplers pushed into the soil at a fast uniform rate produce little disturbance whereas samplers driven into the soil with individual blows (the usual procedure in routine site investigation work in the British Isles) induce considerable disturbance.

Typical thin-walled tube samplers used in the British Isles are shown in Figure 6.4. If the amount of disturbance normally associated with the driven tube sampler is not acceptable, as may be the case when dealing with sensitive soils, then the more sophisticated piston sampler or Swedish foil sampler (Hvorslev, 1949a) should be used and should be pushed into the soil at a uniform fast rate.

In special site investigation work, where it is necessary to study the effects of the macrostructure of the soil on its permeability or undrained shear strength, large-diameter (250 mm) or block samples should be taken.

In general the presence of structural disturbance lowers the shear strength of the soil and increases its compressibility. In the case of tube samples the effects become more pronounced as the degree of overconsolidation of the soil increases (see Section 5.2.2). An interesting study of the effects of sample disturbance was carried out by the BRS (Ward *et al.*, 1959); the undrained shear strength obtained from triaxial tests on 38 mm (1·5 in) diameter driven tube samples was found to be approximately 70% of that obtained from tests on the same size samples hand carved from large block samples. Results of earlier tests carried out by Peck (Terzaghi and Peck, 1967) indicated similar reductions in the strength.

The degree of disturbance of a given soil can be estimated approximately by considering the effects of the change in the state of stress from *in situ* conditions to zero total stress condition. If one assumes that the soil under consideration is elastic, isotropic, and saturated and that the change in the state of stress occurs at a constant water content (i.e. constant volume), then the mean effective stress in the soil during the change must remain constant, i.e. the mean effective stress σ_s induced by the pore water suction acting under the zero total stress condition is equal to the mean effective *in situ* stress σ'_{om}:

$$\sigma'_s = \sigma'_{om} = \frac{\sigma'_{oz} + 2\sigma'_{or}}{3} = \sigma'_{oz}\,\frac{(1 + 2K_0)}{3} \tag{6.1}$$

where σ'_{oz} and σ'_{or} = vertical and horizontal effective overburden stresses
K_0 = coefficient of lateral stress at rest

By estimating the mean effective overburden stress and comparing it with the measured pore water suction (equilibrium stress, Section 5.2.7) the degree of disturbance is estimated: if σ'_{om} is considerably lower than σ'_s then the soil must have dried out; if σ'_{om} is of the same order as σ'_s then the disturbance is small; and if σ'_{om} is considerably greater than σ'_s then the soil must have been disturbed and pore water suction partly destroyed, e.g. by the seepage forces if the level of the water inside the casing was lower than the ground water level outside it.

6.2.3. GROUND WATER OBSERVATIONS

Determination of ground water levels and observations of their fluctuation are amongst the most important site investigation activities. When water is encountered during boring the borehole should be deepened by 0·5 m, the level of water in the borehole should be lowered by 0·5 m (by removing it with a shell) and measurements of the rising water level should be taken every 10 minutes. Boring can be recommenced when the last two readings do not differ by more than 30 mm; the smallest number of observations should not be less than four. The position of the ground water level is determined graphically by plotting the results as shown in Figure 6.5.

Whenever possible the depth of water in boreholes should also be measured during each break in the boring operations; at the beginning of the break, in the middle, and before recommencement of boring.

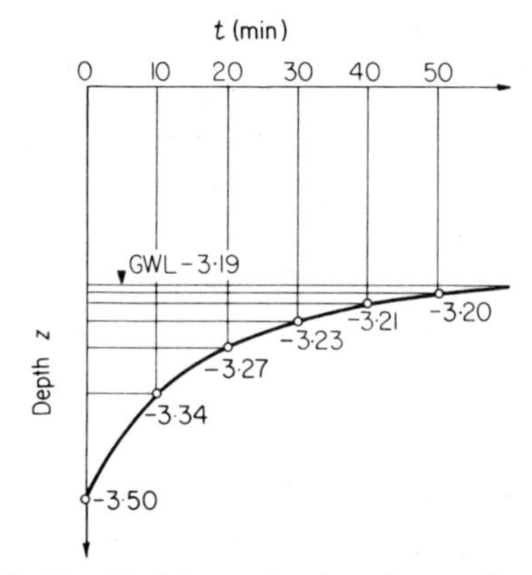

Figure 6.5. Graphical determination of ground water level.

It is very important to measure the levels of water when boring through gravel beds even if observations were previously made in, for example, overlying saturated fine sands. Gravel beds are the most reliable strata for determination of ground water levels.

In the case of design of basements or other underground structures it is recommended that ground water levels and intensity of flow should be determined from observations in trial pits excavated to depths of at least 0·5 m greater than the lowest proposed floor level. Should this not be possible, observation wells or piezometers should be installed in boreholes for the study of ground water levels and *in situ* permeabilities.

In regular ground water conditions ordinary well point type observation wells (stand pipes) are sufficient, whereas in irregular conditions (confined water-bearing strata, Artesian water, etc.) piezometers, installed individually in each permeable layer, may be required. The most common types of piezometers used are shown in Figure 6.6. For detailed treatment of quantitative interpretation of ground water observations refer to Hvorslev (1949b).

From each upper water-bearing stratum (from several boreholes or trial pits) two samples of water should be taken for chemical analysis for determination of its aggressiveness with respect to concrete and steel and, in certain cases, for determination of its suitability for use in mixing of concrete. The containers should be completely filled with water, properly sealed and labelled and, as quickly as possible, dispatched to laboratory for testing.

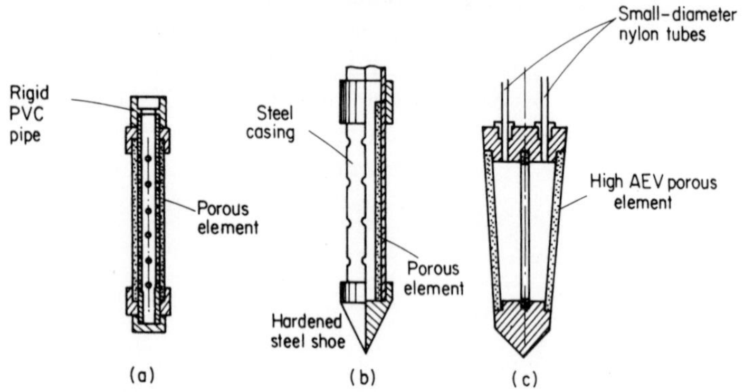

Figure 6.6. Common type of piezometers: (a) Casagrande type tip; (b) Geonor type tip which can be driven into soft soils; (c) Bishop type (high air-entry value) tip for measurement of pore water pressure.

6.3. *In Situ* Tests

6.3.1. STATIC AND DYNAMIC PENETRATION TESTS

Penetration resistance tests or subsurface soundings are used in exploration of erratic soil profiles, in location of bed-rock (or dense strata) and in determination of the state of compaction of cohesionless soils. Because no sounding method is equally suitable under all the soil conditions that may be encountered in the field a great variety of sounding methods have been developed over the last 50 years. Although the penetrometers may differ considerably in design all the sounding methods can be conveniently grouped into two categories: (a) 'dynamic' penetrometers which are driven into the soil by blows of a hammer; (b) 'static' penetrometers which are forced down by steady pressure.

6.3.1.1. Vane/cone Penetrometer (ITB–ZW)

A light dynamic penetrometer developed by Wiłun is shown in Figure 6.7. It can be used with either a conical shoe or a cruciform-type vane blade for subsurface soundings from which penetration-resistance profiles are obtained (Figure 6.8); the latter can also be used for shear vane tests. When driven into cohesionless soils the cone (or vane) penetrometer compacts the surrounding soil by displacement. The penetrometer is driven into the soil by blows from a 22 kg mass falling 250 mm onto a special collar attached to the tubular rods. The number of blows required to drive the penetrometer through every 100 mm is noted. The results are plotted against the depth below ground level as shown in Figure 6.8.

The state of compaction of cohesionless soils is defined according to Table 6.3. It must be remembered that the resistance of soils to both static and dynamic

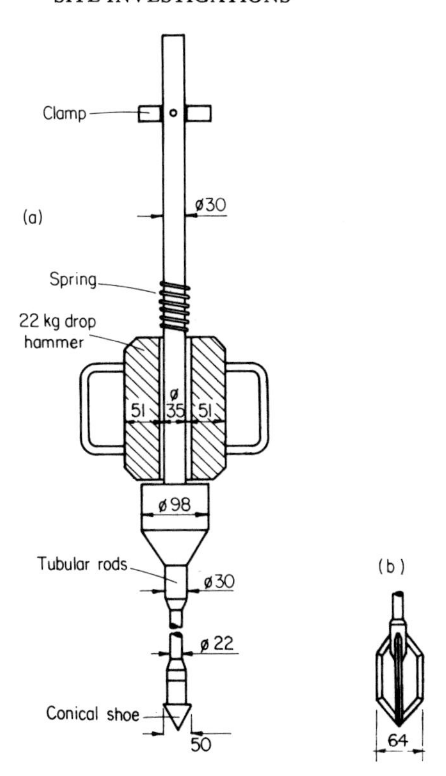

Figure 6.7. Dynamic vane/cone penetrometer (ITB–ZW): (a) conical shoe; (b) vane blade.

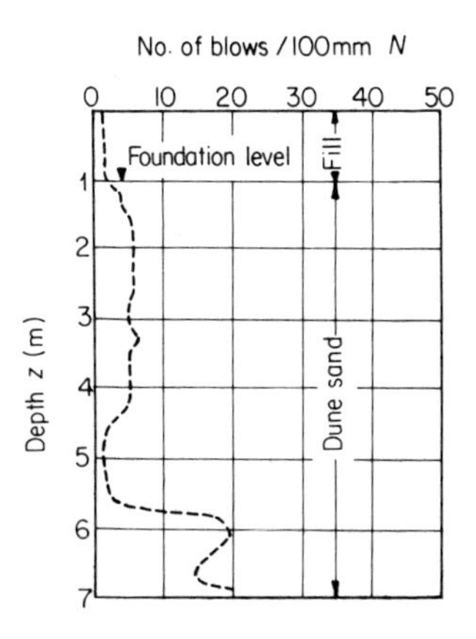

Figure 6.8. Results of subsurface sounding.

penetration depends on a number of factors such as the following: the type of soil tested and its mechanical properties, the mean overburden stress at the test level, the relative position of the ground water table, and the dimensions and shape of the penetrometer.

Therefore the above interpretation which only relates the penetration resistance to the relative density of cohesionless soils or consistency of cohesive soils (see Section 6.3.1.2) is only a simplified generalization which enables one to translate quickly the results of subsurface soundings into meaningful terms and to utilize them for practical purposes; this does not, however, disqualify the whole idea of the subsurface soundings nor the present method of interpretation of the results.

This method of interpretation, being a simplification, must of course incorporate a reasonable margin of safety to allow for the worst possible combination of the above-mentioned factors; it particularly underestimates the state of compaction of sands at shallow depths (low overburden stresses) and below the ground water table.

Table 6.3. State of compaction of cohesionless soils in terms of vane/cone penetrometer results

State of compaction	Density index I_D	No. of blows for 100 mm penetration	
		Conical shoe	Vane blade
very loose	$0{\cdot}00 < I_D \leqslant 0{\cdot}15$	1–4	1–3
loose	$0{\cdot}15 < I_D \leqslant 0{\cdot}33$	4–10	3–6
medium dense	$0{\cdot}33 < I_D \leqslant 0{\cdot}67$	10–30	6–18
dense	$0{\cdot}67 < I_D \leqslant 0{\cdot}85$	30–50	18–30
very dense	$0{\cdot}85 < I_D \leqslant 1{\cdot}00$	>50	>30

To ensure accurate results the hammer must always fall through the full height of 250 mm. In driving the penetrometer through soils below the ground water table the blows should be applied at a rate of one per second; above ground water, at any rate.

When soundings are made in soft soils (organic muds, peats, etc.) the vane blade is used. With the addition of a simple torsion head (Figure 6.9) shear vane tests can be made at any required depth (for interpretation of the shear vane test results see Section 6.3.2).

The resistance to penetration and rotation of the shear vane can be considerably increased by the presence of bent rods. To avoid this the rods are assembled on the ground prior to making soundings and are checked for straightness; any bent rods are rejected. Further resistance to penetration and rotation is encountered owing to adhesion between the soil and the rods when soundings and shear vane test are made in soft cohesive soils. In order to eliminate the influence of these effects on the results, soundings and shear vane tests are repeated in the same hole and at the same depths with a 30 mm diameter (equal to the diameter of rod couplings) conical shoe. The new results are subtracted from the results of the normal soundings with the vane blade. This eliminates the influence of the adhesion on the test results.

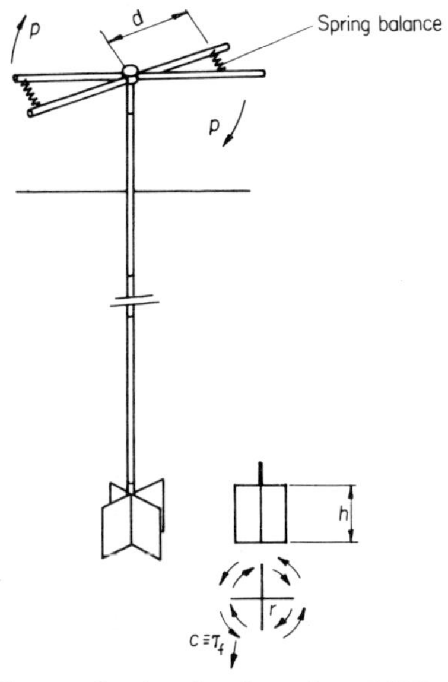

Figure 6.9. Schematic diagram showing the adaptation of (ITB–ZW) penetrometer for determination of the undrained cohesion of soft cohesive soils.

The authors recommend the use of the vane/cone penetrometer for the determination of penetration-resistance profiles (supplemented with shear vane tests in weak soils) of the ground prior to sinking of boreholes. On the basis of these results rational sampling and *in situ* testing programmes can be prepared which will ensure that all soft strata or pockets are thoroughly investigated and that the determined physical and mechanical properties are representative of the encountered soils.

6.3.1.2. *Standard Penetration Test*

The most widely used dynamic method is the standard penetration test, otherwise known as the SPT. The test consists of driving a split-barrel sampler of 50 mm (2·0 in) outside diameter (Figure 6.10) into the soil at the bottom of a cased borehole; in sandy gravels and gravels a solid conical (60°) shoe is fitted to the sampler (Palmer and Stuart, 1957). The sampler is attached to stiff drill rods and is driven into the soil by blows from a 63·5 kg mass (140 lb) falling 0·76 m (30 in) onto the top of the rods fitted with a special drive head. The sampler is initially driven a distance of 150 mm (6 in) below the bottom of the borehole; it is then driven a further distance of 305 mm (12 in) and the number of blows required to drive this distance is termed the *standard penetration value N* (to help in the interpretation of the results the number of blows is recorded every 76 mm (3 in) of the penetration).

The standard penetration test is carried out within the immediate vicinity of the bottom of the borehole, i.e. within the zone of soil which may be disturbed

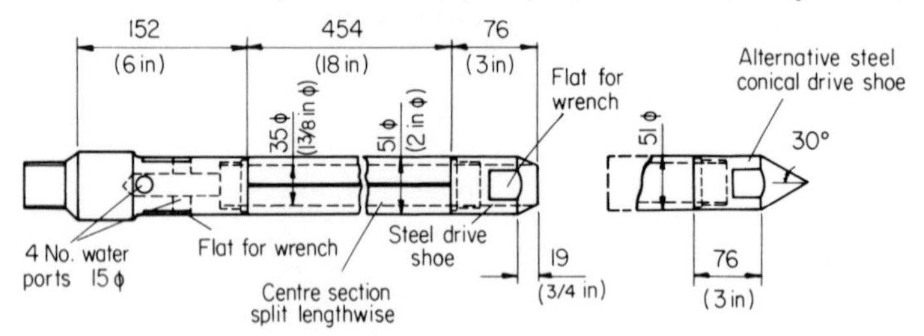

Figure 6.10. Standard penetration test sampler.

by careless boring or, in the case of cohesionless soils, which may be loosened by seepage of water into the borehole. Therefore, in order to obtain reliable results the disturbance of the soil during boring must be kept to the minimum, e.g. by keeping the level of water inside the casing above the ground water level outside it. The test is described in the British Standard 1377 (1967, Test 18).

The state of compaction of cohesionless soils is defined according to Table 6.4 and the consistency of cohesive soils according to Table 6.5. Both these interpretations are subject to the limitations discussed in Section 6.3.1.1; when soundings are made in sands the British Code of Practice 2004 allows rough corrections to be made for the depth effects according to the findings of Gibbs and Holtz (1957).

Table 6.4. State of compaction of cohesionless soils in terms of SPT results

State of compaction	Density index I_D	N value (no. of blows for 305 mm penetration)
very loose	$0{\cdot}00 < I_D \leqslant 0{\cdot}15$	1–4
loose	$0{\cdot}15 < I_D \leqslant 0{\cdot}33$	4–10
medium dense	$0{\cdot}33 < I_D \leqslant 0{\cdot}67$	10–30
dense	$0{\cdot}67 < I_D \leqslant 0{\cdot}85$	30–50
very dense	$0{\cdot}85 < I_D \leqslant 1{\cdot}0$	>50

Table 6.5. Consistency of cohesive soils in terms of SPT results

State of consistency	Consistency index I_c	N value (no. of blows for 305 mm penetration)
very soft to soft	$I_c < 0{\cdot}5$	<4
firm	$0{\cdot}5 \leqslant I_c < 0{\cdot}75$	4–8
stiff	$0{\cdot}75 \leqslant I_c < 1{\cdot}0$	8–15
very stiff or hard	$I_c \geqslant 1{\cdot}0$	15–30
very hard		>30

For tests made in fine sands, silty sands, or silts below the ground water table a correction is usually made for a possible build-up of the pore water pressure: the SPT values (N_0) greater than 15 are modified according to the following expression:

$$N = 15 + \tfrac{1}{2}(N_0 - 15) \tag{6.2}$$

6.3.1.3. *Dutch Cone Penetrometer*

The best known and most widely used static method is the 'Dutch cone' penetrometer test which, as its name implies, was originally developed in Holland for locating dense sand strata into which piles might be driven. The sounding apparatus consists of a probe with a conical point of base area of 1000 mm^2 (35·7 mm diameter) at the end of a rod sliding in a tube of the same external diameter (Figure 6.11). During a test the assembly is pushed into the ground and the cone is advanced a short distance (Figure 6.11(b)) while the force necessary to do this is measured. The tube is then advanced to

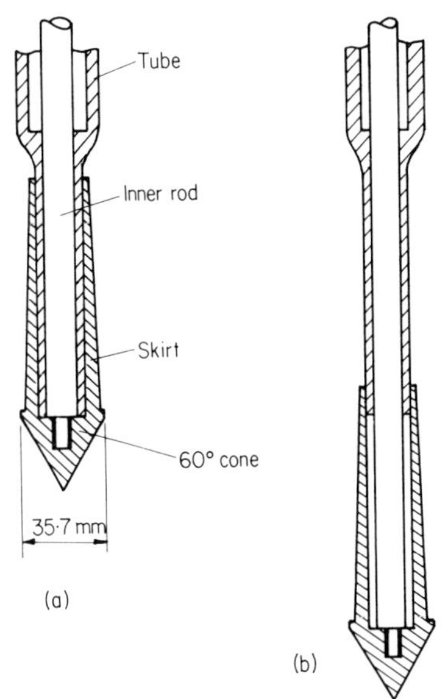

Figure 6.11. Dutch cone penetrometer: (a) as installed in the soil; (b) with the cone advanced.

join the cone and again the force necessary to do this is measured; the readings of cone and tube resistance are usually taken at 200 mm intervals and are plotted against the depth (Figure 6.12). The mechanical systems for measuring the resistance to penetration vary with the manufacturer; most of the modern machines are equipped with automatic load recording devices.

The results of Dutch cone penetrometer tests give continuous profiles of the variation in compaction or consistency of the various encountered soils; this type of information is particularly useful in preparation of geological sections.

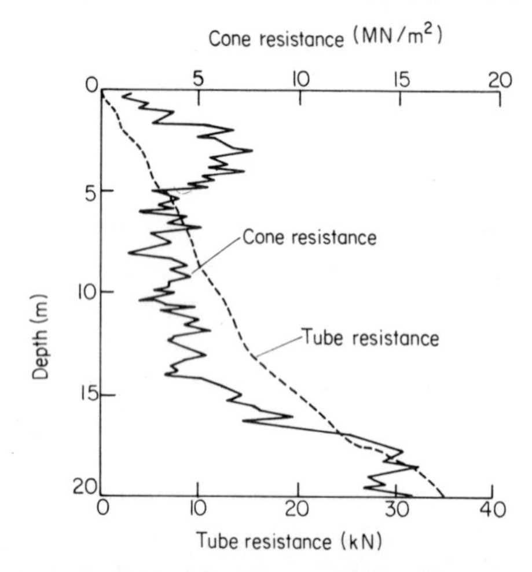

Figure 6.12. Typical Dutch cone penetrometer test results.

In countries such as Belgium and Holland, where soil conditions are fairly uniform, the penetration testing has proved to be a relatively reliable technique and the results are used in evaluation of soil strength, and even compressibility characteristics.

Other forms of both dynamic and static penetrometers are in use and one or other may be favoured in a particular country.

A comprehensive study of the use of penetrometers in soil investigations has been presented by Sanglerat (1965).

6.3.2. SHEAR VANE TEST

In investigation of soft cohesive soils (recent alluvial silts, peat, etc.) the vane penetrometer is frequently used. The vane (Figure 6.13(b)) is pushed or driven into the ground without a boring being made, and at any required depth it is rotated and the maximum torque required to turn the vane is measured; further torsion rods are added and the vane is advanced to the next depth and the test repeated. In order to eliminate the effects on the measured torque of the friction and adhesion between the soil and torsion rods either casing is used (Swedish vane) or the tests are repeated using a conical shoe (of the same diameter as the rods) instead of the vane.

In the British Isles the vane apparatus is generally used in conjunction with the normal type of boring (Figure 6.13(a)) from which samples are taken (Skempton, 1948). Individual tests are usually carried out at a depth of approximately 1·0 m below the bottom of a lined borehole; the torsion rods are provided with guides (top and bottom and at 10 m intervals) to ensure a vertical alignment. Various types of mechanical systems are available for rotating the vane at a constant rate of 0·1° per second and for measuring the torque. The test is covered by the British Standard 1377 (1967, Test 17).

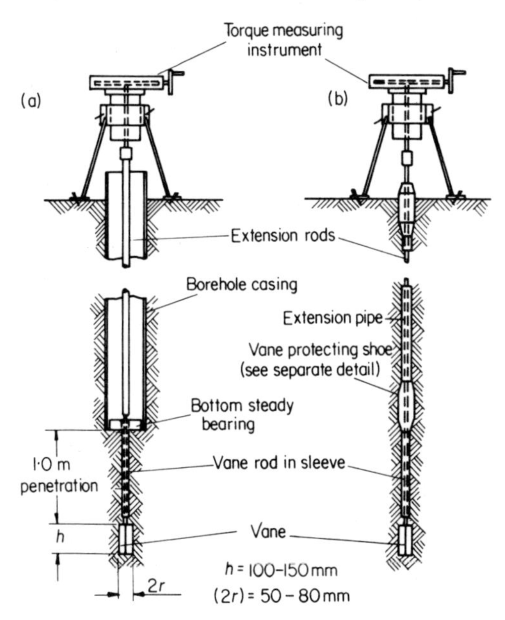

Figure 6.13. Details of shear vane apparatus: (a) borehole vane test; (b) penetration vane test.

In the interpretation of the results it is assumed that for soft cohesive soils sheared in an undrained manner the angle of internal friction $\phi_u = 0°$, i.e. that the shearing resistance is only due to cohesion c_u which is independent of the normal stress. Assuming further that the soil shears along the surface of a cylinder whose diameter and height equal that of the vane and that the shearing resistance across the ends of the cylinder varies linearly, the following expression is obtained:

$$c_u = \tau_f = \frac{M}{2\pi r^2 (h + 2r/3)} \tag{6.3}$$

where M = maximum measured torque
r = radius of the vane
h = height of the vane

Shear vane apparatus also can be used for *in situ* determination of the remoulded strength of soils; on completion of the standard test the vane is rapidly rotated through six or more full rotations and then a further standard rate of shear test is carried out. In this way the sensitivity of clays can be readily determined.

6.3.3. PLATE-LOADING TESTS

In homogeneous soils plate-loading tests are carried out at the proposed foundation level and in at least three different locations in plan. In non-homogeneous (layered) soils tests should be carried out in each layer to a depth (below the foundation level) equal to two widths of the proposed foundation.

In order to obtain reliable and realistic results it is necessary to bed the plate properly to the soil and to prevent (particularly in the case of cohesionless soils) early displacement of the soil from underneath its edges. To achieve

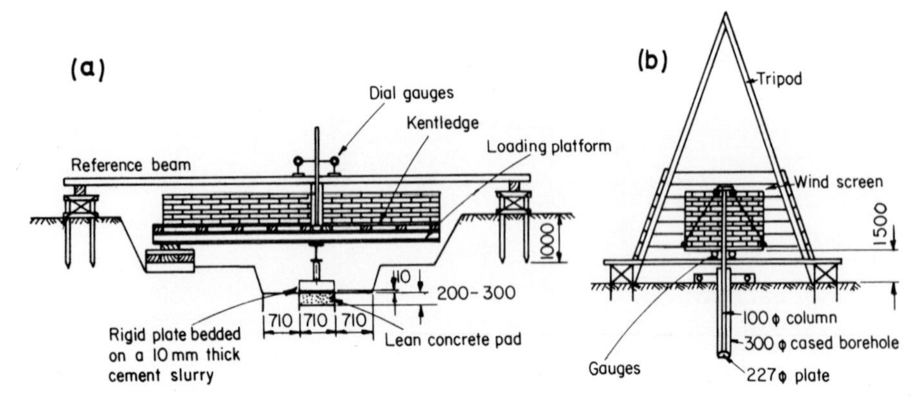

Figure 6.14. Details of plate-loading tests: (a) in a trial pit; (b) in a borehole.

this the authors recommend that for the tests in pits the plate should be positioned on a cast *in situ* concrete pad of thickness equal to 200–300 mm (Figure 6.14). The plate should be bedded to the concrete pad using a 5–10 mm thick layer of cement slurry (or plaster of Paris). In boreholes, generally a special flat-bottom auger is used to clean the bottom of a borehole and the plate is then bedded to the soil by rotating it through several turns. To eliminate the effects of soil disturbance in the immediate vicinity of the bottom of a borehole, particularly in sandy soils, the authors recommend the use of a screw-type plate which is screwed in below the bottom of a borehole to a depth equal to twice its diameter.

On erection of the loading platform and positioning of the dial gauges (fixed to the measuring column and recording settlement with reference to an independent beam) gradual loading of the soil is commenced in steps of 25 or 50 kN/m^2 (0·25 or 0·50 kgf/cm^2, 500 or 1000 lbf/ft^2); the smaller increments are used in the case of soft soils; for this purpose a soil is considered soft when the settlement under the first load increment of 25 kN/m is greater than 0·002B (B = diameter or width of the plate).

After the initial choice has been made, the load increments should be kept constant throughout a test and should be applied at equal time intervals. These intervals should be long enough to ensure that the rate of settlement at the end of each loading stage is less than 0·1 mm in 5 minutes; this means that load increments can be added at intervals of between 40 and 60 minutes in the case of soft (weak) soils, and of not less than 15 to 20 minutes in the case of harder (stronger) soils. The loading is continued until the rate of increase of settlement becomes excessive or settlement equal to 10% of the plate dimension is reached. On application of the last load increment the final load should be maintained

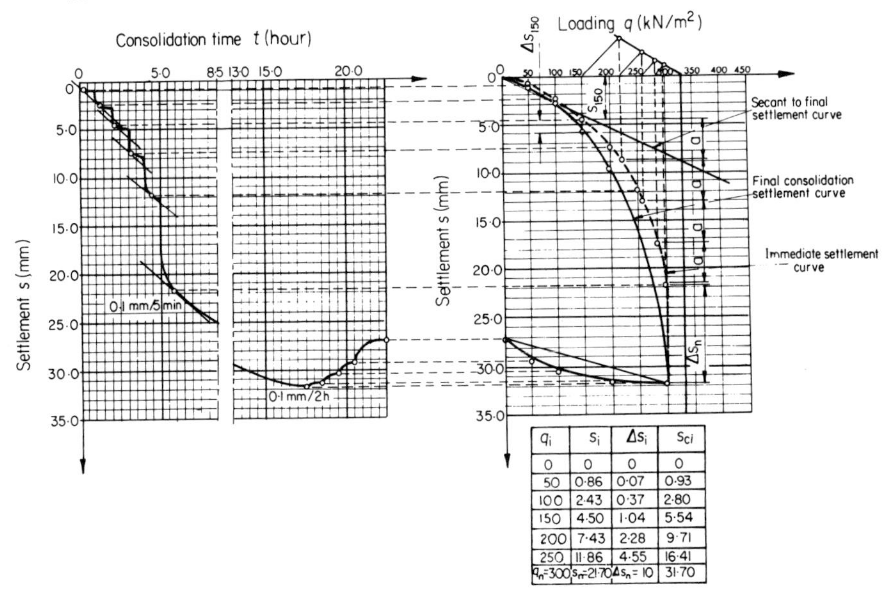

q_i	s_i	Δs_i	s_{ci}
0	0	0	0
50	0·86	0·07	0·93
100	2·43	0·37	2·80
150	4·50	1·04	5·54
200	7·43	2·28	9·71
250	11·86	4·55	16·41
q_n=300	s_n=21·70	Δs_n = 10	31·70

Figure 6.15. Typical results of plate-loading tests: (a) consolidation curves for each loading stage; (b) load–settlement curves.

until the rate of settlement becomes less than 0·1 mm in 2 hours; this can be considered as the end of the primary consolidation stage. On completion of the consolidation stage the plate is unloaded in the same (or doubled) incremental steps and the unloading curve is obtained.

For each loading stage the settlement readings are plotted against time to give a series of consolidation curves as shown in Figure 6.15(a). By drawing a tangent to each consolidation curve, inclined at a slope of 0·1 mm in 5 minutes, tangent points are obtained which are taken to indicate the end of immediate settlement for each loading stage. Using these points the 'immediate' settlement curve is plotted (Figure 6.15(b)) which enables one to determine the ultimate loading, q_{ult}. The final (consolidation) settlement curve is plotted using ordinates evaluated from the following expression:

$$s_{ci} = s_i + \Delta s_i = s_i + \frac{s_i q_i}{s_n q_n} \Delta s_n \qquad (6.4)$$

where s_{ci} = final settlement at loading intensity q_i
s_i = immediate settlement at loading intensity q_i
s_n = immediate settlement at final loading intensity q_n
Δs_n = increase in settlement during the last loading stage from the moment when the rate of settlement is equal to 0·1 mm in 5 minutes until the end of primary consolidation stage (rate of increase 0·1 mm in 2 hours)

The final settlement curve is used in the evaluation of the deformation modulus E_ν. In evaluation of the first loading deformation modulus, E_ν', the slope of the secant from the origin to a point on the curve corresponding to the anticipated stress intensity in a given soil layer is considered; in evaluation of the unloading or reloading deformation moduli, E_ν^u and E_ν'', the slope of the secant covering the whole range of unloading (from q_n to $q = 0$) is considered.

For tests in trial pits the deformation modulus E_ν of the soil in a given layer is evaluated from the following expression:

$$E_\nu = B(1 - \nu^2)\, \omega_{or} \frac{\Delta q}{\Delta s} \qquad (6.5a)$$

where ω_{or} = coefficient corresponding to the shape of the plate: for rigid circular plate $\omega_{or} = 0{\cdot}79$ and for rigid square plate $\omega_{or} = 0{\cdot}88$ (for plates or foundations of other shape and rigidity see Volume 2, Table 2.6)
B = diameter or width of the plate
Δq = increase or decrease in loading
Δs = increase or decrease in settlement due to Δq

For tests in cased boreholes, with the use of a screw-type plate, the deformation modulus E_ν is again evaluated from the slope of the load–settlement curve but correction for the depth of the test is made according to Equation (2.35) of Volume 2:

$$E_\nu = 0{\cdot}79\,(1 - \nu^2)\, d\, \alpha_c \frac{\Delta q}{\Delta s} \qquad (6.5b)$$

where d = diameter of the plate in mm
α_c = value of ratio s_D/s_{max} corresponding to the ratio of the plate diameter d, to its depth below ground level D (for $d/D \leqslant 1/10$, α_c can be taken as 0·5 for all values of Poisson's ratio; for other cases see Figure 2.17 in Volume 2)

Stiffness moduli E' and E'' are determined from Equation (5.9).

Ultimate loading, q_{ult}, can be generally obtained directly from the 'immediate' settlement curve. If for practical purposes it is not possible to reach the ultimate loading, then assuming that the load–settlement curve is of the exponential type a simple geometrical extrapolation can be used to determine q_{ult} (Figure 6.15(b)). Through the tail end of the curve three or four horizontal lines spaced at an equal distance a apart are drawn. Vertical lines are then drawn through the intersection points to intersect the horizontal axis and are produced beyond it. From their point of intersection with the horizontal axis straight lines are drawn at 45° to the horizontal to intersect with the previously drawn vertical lines. The resulting points of intersection are joined with a straight line whose point of intersection with the horizontal axis is taken as the ultimate loading q_{ult}.

The allowable loading of soils should not exceed $q = q_{ult}/2$.

For structures sensitive to differential settlement it is necessary to evaluate settlement and to check if these are within the allowable limits.

In the case of stratified soils or soils of varying consistency, plate-loading tests should be carried out at the upper surface of each stratum or layer of thickness not smaller than three widths of the plate.

6.3.4. PRESSUREMETER TEST

The Menard pressuremeter (Figure 6.16) is used for determination of the *in situ* strength and compressibility characteristics of soils. In the case of soils, in which undisturbed samples cannot easily be extracted for laboratory testing, it is frequently the most economic test for obtaining a sufficiently large number of results for a realistic assessment of their characteristics.

The pressuremeter consists of two basic components: (a) the probe which is situated at the required depth in the borehole, and (b) the pressure volumeter situated at the surface. Pressure is applied through the volumeter to the measuring cell of the probe which then exerts radial stress on the surrounding soil. The volume change of the measuring cell (from which radial strain of the soil can be evaluated) is indicated on the volumeter. The test is performed by raising the pressure in stages at constant short time intervals, volume readings being taken with time at each stage. A pressure–volume change curve (Figure 6.17) is then plotted (taking account of corrections for creep, and instrument volume change effects) and is used for evaluation of soil characteristics.

As can be seen in Figure 6.17 the pressure–volume change curve can be divided into three parts. The initial part of the curve corresponds to the lead-in phase during which the natural earth stress p_0 is being restored. The second part of the curve is almost straight and is taken to correspond to the elastic phase; the pressuremeter deformation modulus E_p is evaluated from the slope of this line. The third part corresponds to the plastic phase during which the

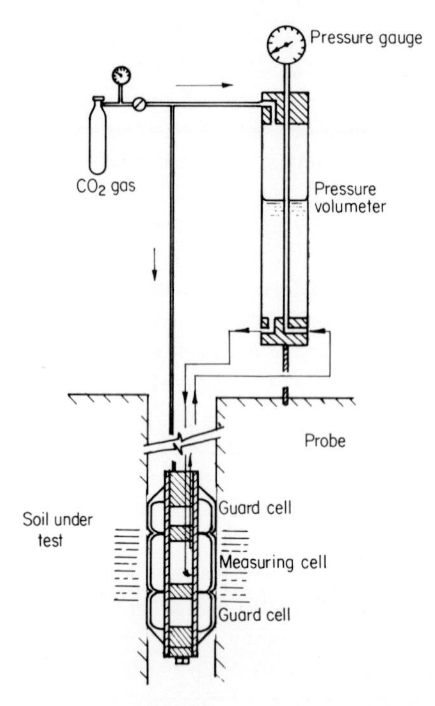

Figure 6.16. Schematic diagram of the Menard pressuremeter.

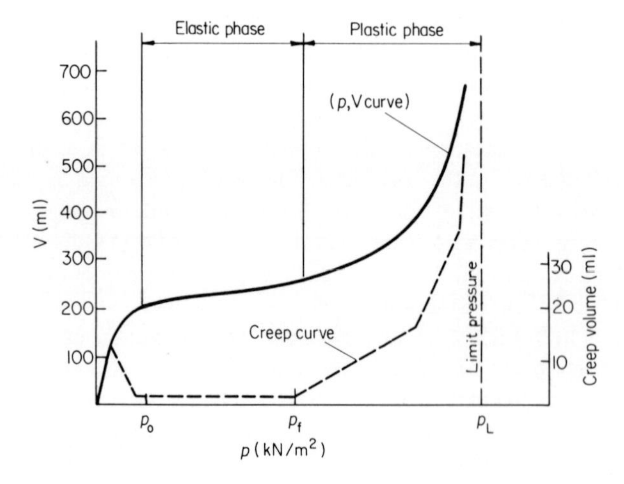

Figure 6.17. Typical pressure–volume change curve.

rate of volume change increases rapidly until the limiting pressure p_L is reached where the curve becomes asymptotic to the vertical. This gives a direct indication of the ultimate bearing capacity of the soil.

The pressuremeter deformation modulus E_p is evaluated from the following expression (Gibson and Anderson, 1961):

$$E_p = 2(1 + \nu) V_0 \frac{\Delta p}{\Delta V} \tag{6.6}$$

where Δp and ΔV = pressure increment and the corresponding change in volume of the cell
V_0 = initial volume of the cell
ν = Poisson's ratio for the soil

For practical purposes Equation (6.6) has been modified to the following form:

$$E_p = K \frac{\Delta p}{\Delta V} \tag{6.6a}$$

where K = a constant for a given probe related to its mean volume V_m in the 'elastic' phase; $K = a + bV_m$, where a and b are constants for a given type of probe

From the knowledge of E_p and of the type of soil being tested the stiffness moduli E' and E'' can be evaluated (Strzelecki and Dudek, 1970):

$$E' = \beta E_p$$

where β = coefficient relating to soil structure, dependent on the type of soil and on E_p/p_L ratio, given in Table 6.6

Table 6.6. Values of coefficient β

Type of soil									
Peats		Clays		Clayey soils, silts		Sands		Stoney soils, gravels	
E_p/p_L	β	E_p/p_L	β	E_p/p_L	β	E_p/p_L	β	E_p/p_L	β
		>16	1·0	>14	1·5	>12	2·0	>10	3·0
	1·0	9–16	1·5	8–14	2·0	7–12	3·0	6–10	4·0
		<9	2·0	<8	2·0	<7	3·0	<6	4·0

Several theories based on assumptions of idealized soils have been developed to facilitate the evaluation of strength parameters from the pressuremeter test results but these do not always give realistic results and further work is required in this field (Komornik *et al.*, 1969).

A series of full-scale tests carried out by Menard at the Centre d'Études Geotechniques de Paris have lead to the formulation of semi-empirical equations which, in certain specific cases, enable one to determine allowable bearing stresses and settlements of foundations directly from the results of pressuremeter tests.

In the case of shallow foundations on homogeneous soils Menard suggests that the allowable bearing stress can be evaluated from the following equation:

$$q_{\mathrm{all}} = q_0 + \frac{k}{3}(p_{\mathrm{L}} - p_0) \tag{6.7}$$

where q_0 = vertical stress in the soil adjacent to the foundation (usually $\gamma_0 D$)

p_{L} and p_0 = horizontal stresses as obtained from test results (Figure 6.17)

k = coefficient from Table 6.8, dependent on the type of soil (Table 6.7), type of foundation, and ratio of the depth of foundation, D, to its width, B

Table 6.7

Type of soil	Consistency/compaction	Class of soil
medium cohesive and cohesive soils (10% < $J \leqslant$ 30%) and silts	plastic (soft to stiff)	I
medium cohesive and cohesive soils (10% < $J \leqslant$ 30%) silts and marls	hard and very hard	II
medium sands	medium dense	II
sands and gravels	dense	III
rocks	soft	III
sands and gravels	very dense	IIIa
rocks	hard	IIIa

Table 6.8. Values of coefficient k

	Square footings				Strip foundations			
Class of soil	I	II	III	IIIa	I	II	III	IIIa
$D/B = 0$	0·8	0·8	0·8	0·8	0·8	0·8	0·8	0·8
0·5	1·3	1·5	1·9	2·1	1·0	1·1	1·2	1·3
1·0	1·6	1·8	2·5	2·8	1·2	1·3	1·4	1·6
1·5	1·8	2·1	3·0	3·3	1·2	1·4	1·6	1·8

In evaluation of settlement of the same foundations Menard introduces an imaginary division of the stressed zone of soil beneath a foundation into two domains: one dominated by spherical compressive stresses and another, outside it, dominated by deviatoric (shear) stresses. The first domain is assumed to be situated immediately beneath the foundation, and to be spherical or cylindrical in shape, and of diameter equal to the width of the foundation.

According to Menard the settlement of the foundation due to volumetric deformations within the spherical domain can be obtained from the following equation:

$$w_s = \frac{2(1-2\nu)}{3E_p} q \, . R \, . \lambda_1 \, . \beta \tag{6.8}$$

where R = radius of the spherical domain ($R = B/2$)

q = applied bearing stress

β = coefficient from Table 6.6

λ_1 = coefficient dependent on the shape of foundation; for circular foundation $\lambda_1 = 1{\cdot}0$ and for rectangular foundations it depends on the $L:B$ ratio as follows:

$L:B$	1·0	2·0	3·0	5·0	20·0
λ_1	1·10	1·20	1·30	1·40	1·50
λ_2	1·12	1·53	1·70	2·14	2·65

For determination of the settlement of the foundation due to shear deformation within the deviatoric domain Menard suggests the following expression:

$$w_d = \frac{(1+\nu)}{3E_p} q R_0 \left(\frac{R}{R_0} \lambda_2\right)^{\beta} \tag{6.9}$$

where R_0 = reference radius ($R_0 = 0{\cdot}30$ m)

λ_2 = coefficient dependent on the shape of foundation; for circular foundation $\lambda_2 = 1{\cdot}0$, for rectangular foundations see Equation (6.8).

The total settlement of the foundation is equal to the sum of w_s and w_d.

The pressuremeter enables one to conduct almost continuous *in situ* testing, throughout a soil profile. The results take into account the heterogeneity of the natural soils and the presence of any discontinuities such as fissures and thus should facilitate a realistic prediction of the soil behaviour under the action of foundation loads.

The results of pressuremeter tests in highly anisotropic soils, such as, for example, soft alluvial deposits interbedded with layers of sand, must be treated with caution. In such a case the sand layers may act as thin stiffening diaphragms which leads to the recording of increased values of E_p and P_L.

6.4. Geotechnical (Site Investigation) Report

Findings of the geological study of a given area (or site), together with the results of exploratory borings, subsurface soundings, *in situ* and laboratory tests, and any other explorations, should be combined into a geotechnical (site investigation) report. Such a report should contain the *geotechnical documentation* of all the investigations and a *description of the site* together *with comments and recommendations* relevant to the design and construction of the proposed works.

A well-documented geotechnical report is a very basic and important source of information in design and planning of construction of all civil engineering works.

6.4.1. GEOTECHNICAL DOCUMENTATION

The geotechnical documentation in a site investigation report should contain the following.

(a) Location plan in 1 : 1000 or 1 : 2000 scale (in British Isles 1 : 1250 or 1 : 2500 scales are common) which should show topographical features of the site in the form of contours, existing and proposed buildings or roads together with position of trial pits or boreholes.

(b) Detailed borehole records.

(c) Geotechnical sections which show individual soil strata and their generalized characteristic properties and ground water levels.

(d) Summary of laboratory and *in situ* tests results.

(e) Summary of chemical analysis of water.

(f) Special information on, for example, mining activities in the area.

Out of the above items the preparation of the geotechnical sections and determination of the generalized characteristic properties of soils in individual strata or layers require a special consideration and are therefore discussed in detail in the following two sections.

6.4.2. PREPARATION OF GEOTECHNICAL SECTIONS

Preparation of geotechnical sections requires a thorough knowledge of geology and considerable amount of practical experience.

In the first instance the soil samples obtained from trial pits and borings are examined and described in terms of colour, particle-size distribution (type of soil), $CaCO_3$ content, and their consistency. If the surface of soil samples stored in boxes has dried out, then the individual lumps are broken and the colour of the inner moist soil is described; on drying, the colour of cohesive soils changes to lighter hue.

The next step involves the description of the type of soil and its structure. Each sample of cohesive soil is broken and its structural characteristics and inclusions (if any) are carefully examined and described according to Section 3.12.3; soils containing cobbles and gravels are of glacial origin (boulder clays), finely laminated soils are usually the varved clays which were deposited in still water impounded in front of a glacier or in glacial lakes, and fissile soils which break easily into parallel-sided fragments are of sedimentary origin (shales).

Cohesionless soils are described according to the grain size and possibly according to their silt content (Table 3.9).

Subsequently, samples with the preserved natural water content (jar or plastic bag samples) are examined and the cohesive soils are described on the basis of the thread-rolling test (Table 3.8).

The most accurate description of the consistency of cohesive soils is obtained from testing of undisturbed samples extracted in tube samplers, both ends of which should be examined.

The results of the macroscopic analysis are superimposed on sections (Figure 6.18) on which the existing ground profiles and positions and centre

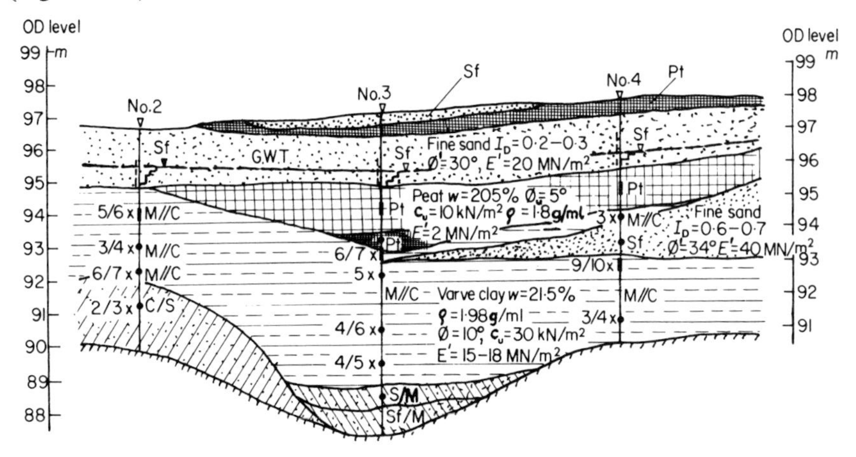

Figure 6.18. Geotechnical section.

lines of boreholes have been drawn (1 : 1000 (1 : 2000) horizontal scale and 1 : 100 (1 : 200) vertical scale). On the right-hand side of the borehole centre lines the type of the soil is indicated according to the abbreviated notation given in Table 6.9 (e.g. M//C, clay with silt laminae) and on the left the number of rolling operations (e.g. 9/10x). The jar or the sealed plastic bag samples are indicated with solid circles and tube samples with solid rectangles. The thickness of strata is indicated with short horizontal lines. Observations of ground water levels are then superimposed.

With the above information superimposed on the section one can proceed with the geological interpretation of the findings and plotting of individual strata. In regions which have been subject to glaciation the arrangement of the basic strata can bc as follows (starting from the bottom): parent rock or soil, sands or gravels deposited prior to glaciation, grey boulder clays (moraine deposits), fluvio-glacial sands and gravels, brown boulder clays (moraine deposits), silts or varved clays (lake deposits), slightly clayey sands or fine and silty sands (Aeolian deposits).

Table 6.9. Abbreviated notation for description of soils
(to be read in conjunction with Tables 3.8 and 3.9)

	Class of soil	Soil types and corresponding symbols					
		Type of soil	Symbol	Type of soil	Symbol	Type of soil	Symbol
cohesive	very cohesive cohesive	sandy clay sand–clay	S/C S–C	clay sand-clay-silt	C S–C–M	silty clay silt–clay	M/C M–C
cohesive	medium cohesive	clayey sand	C/S	sandy clayey silt	S/C/M	clayey silt	C/M
cohesive	slightly cohesive	slightly clayey sand clay with silt laminae	(C)/S M//C	sandy silt silt with clay laminae	S/M C//M	silt	M
	cohesionless	sand coarse sand gravel	S Sc G	silty sand medium sand sandy gravel	M/S Sm S/G	fine sand sand–gravel	Sf S–G
	organic	organic peat	O Pt	organic mud	O/C/M	slightly organic sand	(O)/S

Several different strata of boulder clays can be encountered at any location; varved clays or silts and sands or even peat can be found sandwiched between them.

In certain regions in which the underlying soils were considerably deformed and folded by the weight of the advancing and retreating glaciers, parent soils can often be found in the form of large lenses (erratics) within deep strata of moraine boulder clays. In places where boulder clays were eroded during subsequent formation of river valleys recent alluvial sands can be found directly over the parent rock or soils.

The sequence of strata plotted on longitudinal and transverse sections of any site should correlate with their geological age (Figure 6.18); this should be carefully checked on all the sections. Piezometric levels of ground water in continuous strata of sand should be fairly uniform.

Geotechnical sections should be supplemented with the values of the generalized characteristic properties of soils in the individual strata.

The accuracy of the interpolation of ground conditions in preparation of geotechnical sections depends to a large extent on the spacing between adjacent boreholes; the larger the spacing the greater is the likelihood of marked differences between the interpolated soil conditions and those revealed by subsequent excavation.

6.4.3. DETERMINATION OF GENERALIZED CHARACTERISTIC PROPERTIES OF SOILS

6.4.3.1. Introduction

The fundamental issue in geotechnical engineering is the determination of the generalized characteristic properties of soils for each individual layer (or stratum) of soil on a given site. This information is essential in the design of structural foundations as well as in the consideration of stability of individual structures.

The problem of determination of soil properties for the purpose of design is a very complex one because even the most homogeneous strata in the geological sense, as well as from the point of view of petrological and granulometric composition, exhibit considerable variations in their physical (water content and density) as well as mechanical properties (compressibility and strength). This is due to a number of natural factors.

(1) Non-homogeneity of the parent rock from which the soil was derived.

(2) Variation in the degree of weathering of the mineral constituents of the soil.

(3) Variation in the conditions of sedimentation.

(4) Variation in the conditions of consolidation (at the middle of a stratum and at its boundaries).

(5) The effects of deep frost penetration during glaciation periods (formation of irregular ice lenses).

(6) The effects of tectonic movements or folding under the weight of advancing and retreating glaciers which lead to the formation of structural discontinuities along the slip planes (slickensides).

Soil properties can also be considerably altered by the following.

(a) Extraction of samples from the bottom of a borehole under conditions of seepage.

(b) Disturbance of soil structure during sampling.

(c) Changes in water content during storage of samples: migration of water from less to more cohesive layers of laminated soils or drying out.

Obviously one must not allow changes to take place in the natural properties of soils due to bad sampling or storage; any affected samples should be completely eliminated from the considerations leading to the determination of the characteristic properties of soils.

Particular attention should be given to the sampling of soils containing slip planes (slickensides), and finely laminated soils. The samples should be extracted in such a manner that the planes of structural discontinuity should be inclined at the same angle to the planes of shear failure in the laboratory apparatus as they would be in a given layer in natural conditions beneath a foundation or in a slope.

In general one should attempt to determine the mechanical properties of soils in conditions which as closely as possible resemble the natural conditions in the ground beneath a foundation or in the slope of a cutting.

Taking the above into consideration one finds that even after the elimination of the effects of sample disturbance there still is a considerable scatter of the results which is partly due to the non-homogeneity of the soil in a given layer and partly to the inaccuracies of laboratory testing. Therefore in order to determine significant characteristic properties of the soil in a given layer it is necessary to use an appropriate method of analysis of the results.

6.4.3.2. Statistical Methods

There are a number of statistical methods which can be used in the analysis of a set of test results with the object of determination of the significant characteristic properties of the whole mass of the soil in a given layer.

A relatively simple method, which is sufficiently accurate for most practical purposes is based on the determination of the best estimates of the arithmetic mean $\overline{x}$ and standard deviation σ_{est} for the population from a given set of test results and on the evaluation of confidence limits x_α within which one is $\alpha\%$ certain that the population mean (i.e. the characteristic property) of the required soil property is situated. The equations used in this analysis are

$$\overline{x} = \frac{\sum_{i=1}^{N} x_i}{N} \tag{6.10}$$

$$\sigma_{est} = \sqrt{\left\{\frac{\sum_{i=1}^{N} (x_i - \overline{x})^2}{N-1}\right\}} \tag{6.11}$$

$$x_\alpha = \overline{x} \pm t_\alpha \frac{\sigma_{est}}{\sqrt{N}} \tag{6.12}$$

where $\overline{x}$ = arithmetic mean of the given set of test results or the best estimate of the arithmetic mean for the population

x_i = single test result (e.g. water content or liquid limit, etc.)

N = number of test results in the set

σ_{est} = the best estimate of the standard deviation for the population

t_α = student's t value obtained from Table 6.10 for the required level of confidence α

α = level of confidence, i.e. the required degree of certainty that the population mean lies within the stated confidence limits

The above equations refer to a population (in this case the whole mass of soil within an isolated layer) of normal (Gaussian) distribution. It can be assumed on the basis of up-to-date practical experience that geotechnical properties of soils follow the normal distribution. This can, in each case, be verified by plotting the results in the form shown in Figure 6.19; values of test results are plotted on the horizontal axis (linear scale) and the corresponding number of test results expressed in percentages are plotted cumulatively on the vertical axis (Gaussian scale). If the resulting plot approaches a straight line, then the distribution of the results approaches normal distribution; a sudden change in the slope of the line indicates that samples from two different formations have been included whereas a general scatter of the points indicates inadequate size of the set.

Table 6.10. Student's t values

Number N	Confidence level α			
	90%	95%	99%	99·9%
1	6·31	12·71	63·66	636·62
2	2·92	4·30	6·92	31·60
3	2·35	3·18	5·84	12·92
4	2·13	2·78	4·60	8·61
5	2·02	2·57	4·03	6·87
6	1·94	2·45	3·70	5·96
7	1·90	2·37	3·50	5·40
8	1·88	2·30	3·36	5·04
9	1·83	2·26	3·25	4·78
10	1·81	2·23	3·17	4·59
11	1·80	2·20	3·11	4·44
12	1·78	2·18	3·06	4·32
13	1·77	2·16	3·01	4·22
14	1·76	2·14	2·98	4·14
15	1·75	2·13	2·95	4·07
16	1·75	2·12	2·92	4·02
17	1·74	2·11	2·90	3·97
18	1·73	2·10	2·88	3·92
19	1·73	2·09	2·86	3·88
20	1·72	2·09	2·85	3·85
21	1·72	2·08	2·83	3·82
22	1·72	2·07	2·82	3·79
23	1·71	2·07	2·81	3·77
24	1·71	2·06	2·80	3·75
25	1·71	2·06	2·79	3·72
26	1·71	2·06	2·78	3·71
27	1·70	2·05	2·77	3·69
28	1·70	2·05	2·76	3·67
29	1·70	2·05	2·76	3·66
30	1·70	2·04	2·75	3·65
40	1·68	2·02	2·70	3·55
50	1·67	2·00	2·66	3·46
120	1·66	1·98	2·62	3·37
∞	1·65	1·96	2·58	3·29

The evaluation of $\bar{x}$ and σ_{est} on the basis of Equations (6.10) and (6.11) is tedious (unless computer facilities are available) and therefore the following approximate, but still sufficiently accurate, simpler and quicker method can be used.

The method is based on the technique of grouping of the results in equally spaced intervals and on the use of the following equations:

$$\bar{x} = \frac{\sum xn}{N} \tag{6.13}$$

or

$$\bar{x} = x_a + \frac{a}{N} D_a \tag{6.14}$$

and

$$\sigma_{est} = \pm \sqrt{\left\{\frac{a^2}{N} S - \left(\frac{a}{N} D_a\right)^2\right\}} \tag{6.15}$$

where x = central value for a group
n = number of test results in a given group
$\bar{x}_a$ = convenient assumed arithmetic mean of the complete set
a = group interval
D_a = difference of sums S_1 and S_2 (see example in Table 6.11)
$S = S_a + 2S_a'$ (see example in Table 6.11)

The results of water content tests given in Figure 6.19 will now be analysed to illustrate the above method.

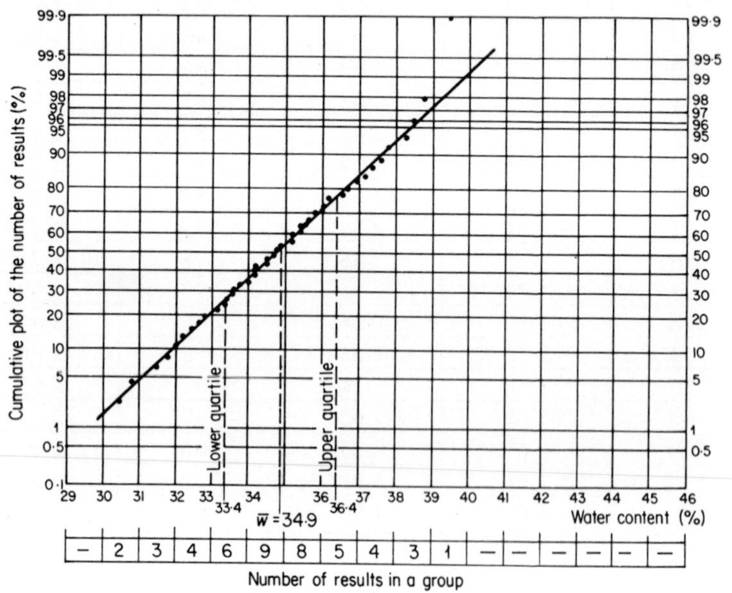

Figure 6.19. Distribution of results of water content tests plotted on Gaussian scale.

Example. The test results are grouped in 1·0% water content intervals and are tabulated as shown in Table 6.11 (columns 1, 2, and 3), a convenient arithmetic mean $\bar{x}_a$ is assumed on the basis of Figure 6.19 and calculations in columns 4, 5, and 6 are completed.

The arithmetic mean of the water contents can be determined according to Equation (6.13):

$$\bar{x} = \frac{\sum xn}{N} = \frac{1468{\cdot}5}{45} = 34{\cdot}86 \approx 34{\cdot}9$$

or according to Equation (6.14)

$$\bar{x} = 34{\cdot}5 + \frac{1{\cdot}0}{45} \times 16 = 34{\cdot}5 + 0{\cdot}36 \approx 34{\cdot}9$$

The standard deviation is evaluated from Equation (6.15):

$$\sigma_{\text{est}} = \pm \sqrt{\left\{\frac{1{\cdot}0^2}{45} \times 218 - \left(\frac{1{\cdot}0}{45} \times 16\right)^2\right\}} = \pm\sqrt{4{\cdot}72} = \pm\, 2{\cdot}17$$

When the standard deviation is expressed as a percentage of the mean, one obtains the coefficient of variation:

$$V = \frac{\sigma_{\text{est}}}{\bar{x}} \times 100 \qquad (6.16)$$

For the considered example $V = 6{\cdot}2\%$. In general, values of V below 10 indicate very good uniformity within a given set whereas values between 10–13, 13–16, 16–20, and above 20 indicate, respectively, good, average, low, and very poor uniformity.

The percentage error in evaluation of the population mean for the various degrees of confidence can be taken as equal to

$$\rho_x = \frac{t_\alpha\, \sigma_{\text{est}}}{\bar{x}\sqrt{N}} \times 100 \qquad (6.17)$$

For the considered example

(1) for $\alpha = 95\%$ $\qquad \rho_x = \dfrac{2{\cdot}01 \times 2{\cdot}17}{34{\cdot}9\sqrt{45}} \times 100 = 1{\cdot}9\%$

(2) for $\alpha = 99{\cdot}9\%$ $\qquad \rho_x = \dfrac{3{\cdot}50 \times 2{\cdot}17}{34{\cdot}9\sqrt{45}} \times 100 = 3{\cdot}3\%$

i.e. there is a 5% (1 in 20) chance of the error being 1·9% and only 0·1% (1 in 1000) chance of there being a 3·3% error.

This indicates considerable accuracy in determination of the population arithmetic means of the considered property (water content).

Table 6.11. Determination of mean water content by tabular method (example)

Group limits (Figure 6.19)	Central value for a group x	Number in a group n	Assumed mean $\bar{x}_a = 34{\cdot}5$ — Sum of numbers in groups: First	Second	Product $x.n$
1	2	3	4	5	6
30·0–31·0	30·5	2	2 (S_2)	2 (S_2')	61·0
31·0–32·0	31·5	3	5 (S_2)	7 (S_2')	94·5
32·0–33·0	32·5	4	9 (S_2)	16 (S_2')	130·0
33·0–34·0	33·5	6	15 (S_2)	–	201·0
34·0–35·0	34·5	9	Assumed mean		310·5
35·0–36·0	35·5	8	21 (S_1)	–	284·0
36·0–37·0	36·5	5	13 (S_1)	26 (S_1')	182·5
37·0–38·0	37·5	4	8 (S_1)	13 (S_1')	150·0
38·0–39·0	38·5	3	4 (S_1)	5 (S_1')	115·5
39·0–40·0	39·5	1	1 (S_1)	1 (S_1')	39·5
		$N = 45$	$S_1 = 21 + 13 + 8 + 4 + 1 = 47$ $S_2 = 2 + 5 + 9 + 15 = 31$ $S_a = 47 + 31 = 78$ $D_a = 47 - 31 = 16$ $S = 78 + 2 \times 70 = 218$	$S_1' = 26 + 13 + 5 + 1 = 45$ $S_2' = 2 + 7 + 16 = 25$ $S_a' = 45 + 25 = 70$ $D_a' = 45 - 25 = 20$	1468·5

Note. Figures in column 4 are obtained by successive summation of figures in column 3, downwards and upwards to the assumed mean; similarly figures in column 5 are obtained by successive summation of figures in column 4.

The confidence limits x_α are evaluated from Equation (6.12). For a given number N of test results in a set, t_α values corresponding to the required level of confidence ($\alpha = 90, 95, 99$, and 99·9%) are taken from Table 6.10 and on substitution in the equation the upper and lower confidence limits x_α are obtained (Table 6.12).

Table 6.12. Lower and upper confidence limits x_α

α	t_α	$t_\alpha \frac{\sigma_{est}}{\sqrt{N}}$	Values of x_α: Lower	Upper
90	1·68	0·5	34·4	35·4
95	2·02	0·75	34·2	35·6
99	2·69	0·9	34·0	35·8
99·9	3·53	1·1	33·8	36·0

Taking into consideration the low level of accuracy associated with geotechnical testing and the considerable natural variation of soils, even in apparently homogeneous layers, one has to obtain a large number of test results in order to determine the generalized characteristic properties with a reasonable degree of accuracy. At the same time, however, it is possible to use simpler methods of analysis in determination of these properties.

A method based on determination of the guaranteed average minimum and maximum values according to Equations (6.18) and (6.19) can be used for approximate estimation of the confidence limits. After 10% of the lowest or highest results in a given set have been rejected

$$x_{av\,min} = \frac{\overline{x} + x_{min}}{2} \quad (6.18)$$

$$x_{av\,max} = \frac{\overline{x} + x_{max}}{2} \quad (6.19)$$

In the considered example one obtains, from Figure 6.19, $x_{min} = 31{\cdot}9$ and $x_{max} = 37.7$, and therefore the average minimum and maximum values

$$x_{av\,min} = \frac{34{\cdot}9 + 31{\cdot}9}{2} = 33{\cdot}4\%$$

$$x_{av\,max} = \frac{34{\cdot}9 + 37{\cdot}7}{2} = 36{\cdot}3\%$$

There is also a possibility of considering the lower and upper quartiles as approximate limits of the confidence interval; with sufficient accuracy for practical purposes these values can be determined from Figure 6.19, by rejecting 25% of the test results from each end of the plot; the two ends of the shortened cumulative plot correspond to the quartiles. The lower quartile in Figure 6.19 is 33·4% and the upper 36·4%; the upper significant characteristic value of the water content would then be taken as $w_{rcp} = 36{\cdot}4\%$.

As can be seen from the considered example the values of water content corresponding to the confidence limits obtained by different methods fall within a fairly narrow band, the intermediate values being those obtained by the method of quartiles. Considering the simplicity of application of this method and of its clear physical interpretation it is recommended as the most effective method for general use.

In the case of investigations of special importance and when V is greater than 16 the authors recommend the use of the 'confidence limits' method; otherwise the simple 'quartiles' method is considered to be sufficiently accurate (Table 6.12).

The above-described methods of evaluation of the confidence limits for a given population can be applied in determination of any geotechnical property, provided a sufficiently large number of results is available from tests carried out on samples which are representative of the soil in a given layer.

This condition can be satisfied in the case of water content tests which are relatively simple and are usually carried out in sufficiently large numbers. In the case of the more time- and work-consuming investigations (e.g. determination of compressibility or shear strength) execution of a sufficiently large number of tests is very difficult and an analysis of results of a small number of tests may be subject to considerable accidental errors because the samples may not be sufficiently representative of the soil in a given layer.

In connection with the above the authors recommend the use of the graphical correlation method.

6.4.3.3. Graphical Correlation Method

The method is based on a graphical summary of results of tests carried out to determine various soil properties, with reference to a 'leading' property (Figure 6.20); it is suggested that for cohesive soils the water content is taken as the leading property, and for cohesionless soils the density index. When all the results are plotted correlation curves can be drawn. Frequently the plotted results are not evenly distributed along the reference line (horizontal axis); it is therefore suggested that the reference line is divided into equal length sections (groups) so that within each group at least two results are present; weighted average values are then evaluated for each group and the new points are used for plotting correlation curves (or straight lines).

The ordinate corresponding to the point of intersection of the correlation curve of the considered property with a vertical line drawn through the appropriate confidence limit of the leading property is taken as the significant characteristic value of the property.

The recommended method has the following basic attributes.

(1) It forms a synthesis of the complete range of investigations in a given layer.

(2) It enables one to reject accidental errors or mistakes.

(3) It gives a good insight into relationships between individual properties and enables one to carry out a comprehensive analysis of the considered geotechnical problem.

The graphical correlation method is considered to be sufficiently accurate for most practical purposes. If in certain cases the use of numerical correlation methods is considered necessary then, again, it is suggested that as the first step in the analysis the results are divided into groups and then the correlation is carried out using the weighted group averages. This is justified in the physical sense, because it enables one to eliminate the effects of accidental accumulation of results in several groups which otherwise might have influenced the results of the correlation analysis by giving incorrect coefficients of regression; the grouping also simplifies and reduces numerical work.

In cases when tests of different accuracy are used in determination of a particular soil property (e.g. determination of compressibility of soil from results of oedometer and plate-loading tests) the correlation curves should be drawn separately for each method of testing (Figure 6.20); the lower correlation

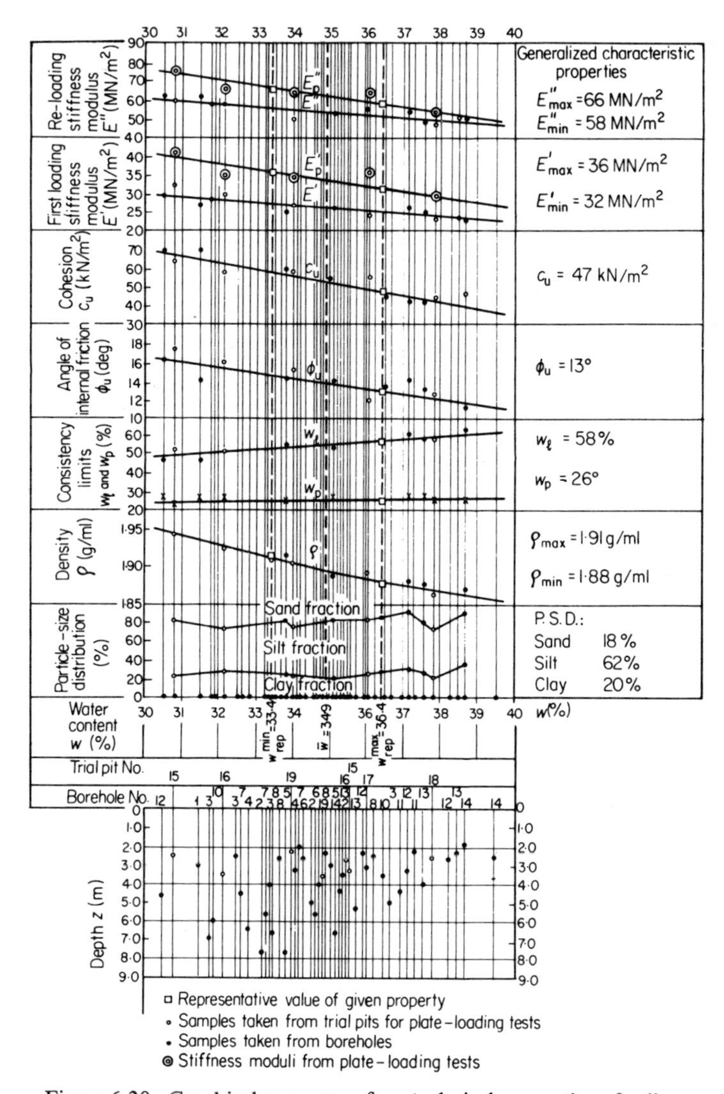

Figure 6.20. Graphical summary of geotechnical properties of soil.

curve for the less accurate oedometer method is obtained on the basis of a large number of results and it can be used for comparative reference purposes in plotting the higher correlation curve for the smaller number of the more accurate results of the plate-loading tests. The latter correlation curve should be used in determination of the significant characteristic properties.

The correlation curve relating stiffness moduli to water content is used for determination of the upper and lower values of the characteristic stiffness moduli E'_{max} and E'_{min} for a given layer; these are obtained by plotting vertical lines through the lower and upper water content confidence limits (see Figure 6.20).

The obtained characteristic stiffness moduli E'_{max} and E'_{min} can be used in evaluation of the expected total and differential settlements of foundations of a given building. In connection with this, one should verify whether the most compressible zones of soil (having $E' < E'_{min}$) are not, for example, situated under the central part of the building while the less compressible (having $E' > E'_{max}$) under the end walls, or vice versa.

Shear strength correlation curves (ϕ_u, w) and (c_u, w) curves, are usually used only for the determination of the lower significant characteristic values $\phi_{u\,min}$

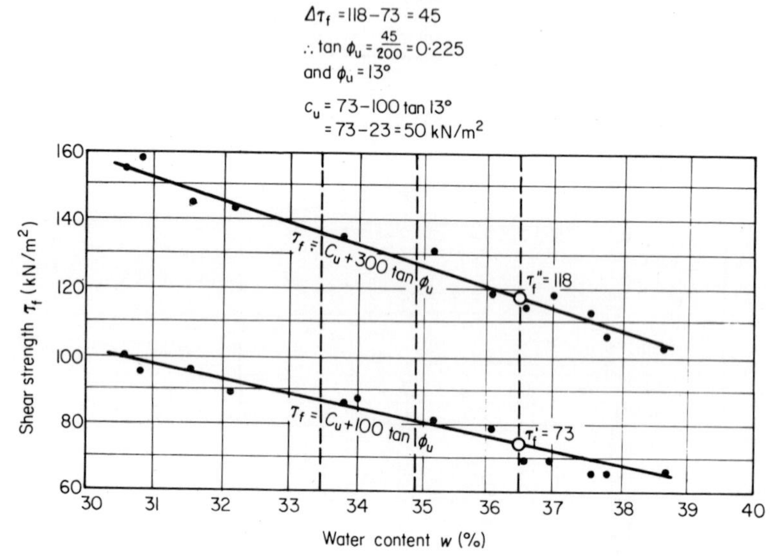

Figure 6.21. Graphical determination of significant characteristic values of c_u and ϕ_u using shear strength τ_f.

and $c_{u\,min}$, i.e. the values corresponding to the upper water content confidence limit.

If the scatter of ϕ_u and c_u values is greater than that in Figure 6.20, then a more accurate method can be used which is based on the consideration of the shear strength $\tau_f = c_u + \sigma_n \tan \phi_u$. By assuming two values of σ_n (one, for example, 100 kN/m^2 smaller and the second 100 kN/m^2 greater than the expected value of σ_n) the corresponding values of τ_f are calculated and are plotted as shown in Figure 6.21. Correlation curves are drawn and the upper and lower significant values $\tau_{f\,max}$ and $\tau_{f\,min}$, corresponding to the lower and upper water content confidence limits, are used in evaluation of the characteristic values of $\phi_{u\,min}$ and $c_{u\,min}$.

As can be seen, the strength parameters obtained from Figure 6.20 are $\phi_{u\,min} = 13^\circ$ and $c_{u\,min} = 47\ kN/m^2$ and, from Figure 6.21, $\phi_{u\,min} = 13^\circ$ and $c_{u\,min} = 50\ kN/m^2$.

6.4.3.4. Generalized Characteristic Soil Properties and Factors of Safety

The values of factors of safety at present used in geotechnical designs have been established on an empirical basis. For any specific problem wide ranges of values are frequently quoted, as, for example, in the case of determination of allowable bearing stresses where factors of safety between 2 and 3 are used, when the ultimate bearing capacity is considered, and wide ranges of allowable settlements or angular distortions (e.g. 1 : 200–1 : 400) are used, when settlements are taken into consideration; no reference is made with regards to the accuracy of determination of the geotechnical properties of the soil.

Undoubtedly the large values of the factors of safety used possess a sufficiently large margin of safety to cover for the inaccuracies associated with the methods of design and soil parameters.

There is therefore a possibility of linking the values of factors of safety with the accuracy in determination of soil properties.

A general principle can be adopted that, in cases where more accurate methods of testing and determination of generalized characteristic soil properties are used, lower factors of safety are acceptable.

In such cases a comprehensive analysis of the ground conditions and test results is essential; a graphical summary of all tests results of the type shown in Figure 6.20, at the bottom of which borehole and trial pit reference numbers and depths of samples are plotted, can be very useful.

This enables one to obtain a clear three-dimensional picture of the ground conditions which will indicate, for example, whether the weaker soils in a given layer are associated with a particular area on the site or whether the lower values are part of the natural scatter of results in a given set.

In the case of design of important structures the graphical summary of test results should form an integral part of the site investigation report.

6.4.4. DESCRIPTION OF SITE, COMMENTS, AND RECOMMENDATIONS

The description of the site, comments, and recommendations relevant to the proposed works should cover the following.

(a) Topography of the site.

(b) Geological structure, i.e. indication of the origin of the soils and their geological characteristics.

(c) Ground water conditions with indication of possible fluctuations of the ground water table and present and probable differences in water levels in adjoining surface reservoirs.

(d) Description of soil strata: the type and consistency of soils, their physical and mechanical properties and indication of the range of probable bearing stresses, their frost susceptibility (mainly for road construction), their suitability for construction of embankments and indication of the local sources of road materials.

(e) A discussion of the possibility of landslides in natural slopes, new cuttings, and excavations.

(f) The type of construction and state of existing buildings and roads.

(g) A discussion of any special problems associated with the type of the proposed works.

7

Nomograms for Determination of Physical Properties of Soils

7.1. Introduction

The knowledge of a wide range of physical properties of soils is necessary in the investigation of their shear strength and compressibility characteristics in the laboratories as well as in the actual calculations leading to the determination of safe bearing stresses, thrust on retaining walls, or the stability of slopes. The following properties belong to this range.

(1) ρ_d, dry density of soils.
(2) ρ_{sat}, density of fully saturated soil.
(3) n, porosity of soil.
(4) e, voids ratio.
(5) w_{sat}, saturation water content.
(6) S_r, degree of saturation.

All the above properties can be computed from the knowledge of the following basic properties.

(a) G_s, specific gravity of solid particles.
(b) ρ, natural density of soil.
(c) w, natural water content.

The computations are, however, tedious and time consuming and the use of appropriate nomograms can be of considerable assistance.

7.2. Description of Nomograms

Five nomograms have been constructed for the following individual values of the specific gravity of the solid particles: $G_s = 2{\cdot}65$, $2{\cdot}67$, $2{\cdot}69$, $2{\cdot}72$, and $2{\cdot}75$. For soils having particles of specific gravity between these figures the nomogram with G_s closest to the actual value is used. Water contents w in percentages are plotted on the horizontal axes of the nomograms while the natural densities ρ are plotted on the left-hand vertical axis.

The sloping lines (rising from left to right) represent values of porosity n and voids ratio e as functions of the coordinates w and ρ. The points of intersection of such a line with the ordinate axis and with the full saturation line ($S_r = 1{\cdot}0$) enable one to read off directly the values of dry density ρ_d and saturated density ρ_{sat} of any soil represented by a point situated on the line.

The full saturation densities ρ_{sat} are marked on the right-hand vertical axis.

The sloping lines (rising from right to left) give the degree of saturation S_r of the soil.

7.3. Accuracy of Results

For the normal size nomograms (A4) and in the case of their careful use the errors do not exceed 1% of the determined values. The correct nomogram should be selected for the given value of G_s. When it is necessary to approximate, then, for a difference in $\Delta\rho$ of less than or equal to 0·02, the errors do not exceed 2%. It is important, however, to use the one selected nomogram for determination of all the properties of a given soil. For example, if the soil has a specific gravity of $G_s = 2{\cdot}66$ and nomogram $G_s = 2{\cdot}65$ has been used then, when dealing with the same soil on subsequent occasions, the same nomogram should be used and not the one for $G_s = 2{\cdot}67$.

7.4. Instructions for the Use of Nomograms

In Figure 7.1 the key diagram to the use of the nomograms is shown. For a given G_s the following points are marked on the appropriate nomogram.

(1) Point A, corresponding to the natural water content w, is marked on the abscissa.

(2) Point B, corresponding to the natural density ρ is marked on the ordinate.

(3) Straight lines drawn through these two points parallel to the axes determine point M.

(4) If the point M falls on one of the sloping lines $n(e)$, then the figures written above the line give the corresponding values of n and e; if, as shown on the key diagram, the point M falls between two sloping lines, then the corresponding values of n and e are obtained from linear interpolation; the same applies to S_r lines.

(5) Sloping line drawn through the point M parallel to the n lines intersects the left-hand axis at point N and the full saturation line w_{sat} at point O.

(6) The reading on the ordinate corresponding to N gives the dry density ρ_d.

(7) A horizontal straight line through point O intersects the right-hand axis at P giving the value of ρ_{sat}.

(8) A straight line drawn vertically through O intersects the upper horizontal axis at Q and determines the value of w_{sat}.

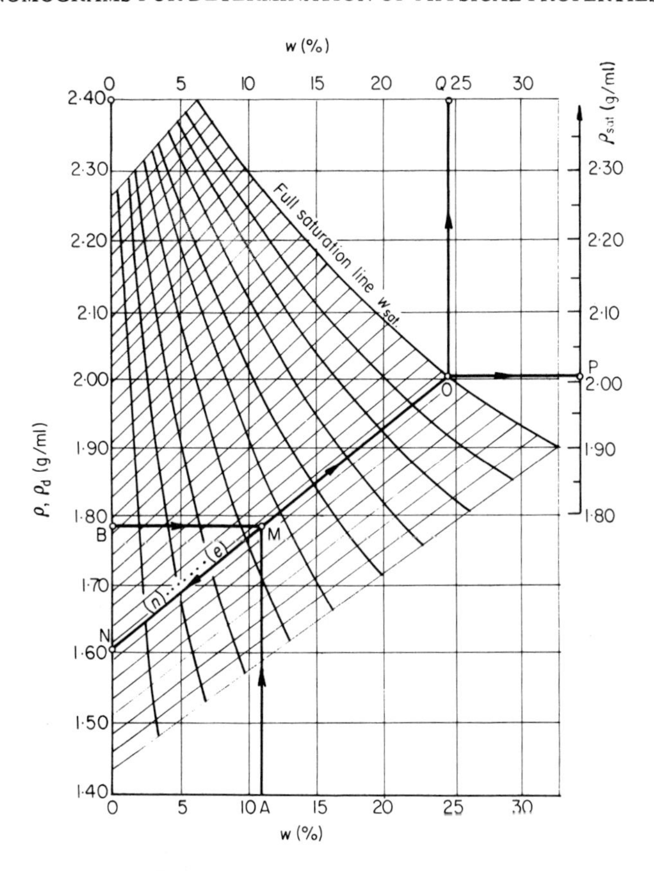

Figure 7.1. Key diagram to the use of nomograms.

The presented nomograms are shown

in Figure 7.2 for $G_s = 2{\cdot}65$
in Figure 7.3 for $G_s = 2{\cdot}67$
in Figure 7.4 for $G_s = 2{\cdot}69$
in Figure 7.5 for $G_s = 2{\cdot}72$
in Figure 7·6 for $G_s = 2{\cdot}75$

Example. Data: $G_s = 2{\cdot}65$, $\rho = 1{\cdot}845$ g/ml, $w = 10{\cdot}2\%$. Determine n, e, ρ_d, ρ_{sat}, w_{sat}, S_r, and γ_{sub}.

From Figure 7.2 we read off $n = 0{\cdot}368$, $e = 0{\cdot}583$, $\rho_d = 1{\cdot}674$ g/ml, $\rho_{sat} = 2{\cdot}046$, $w_{sat} = 22{\cdot}2\%$, $S_r = 0{\cdot}47$, $\gamma_{sub} = 9{\cdot}8\,(2{\cdot}05 - 1{\cdot}0) = 10{\cdot}3$ kN/m^3.

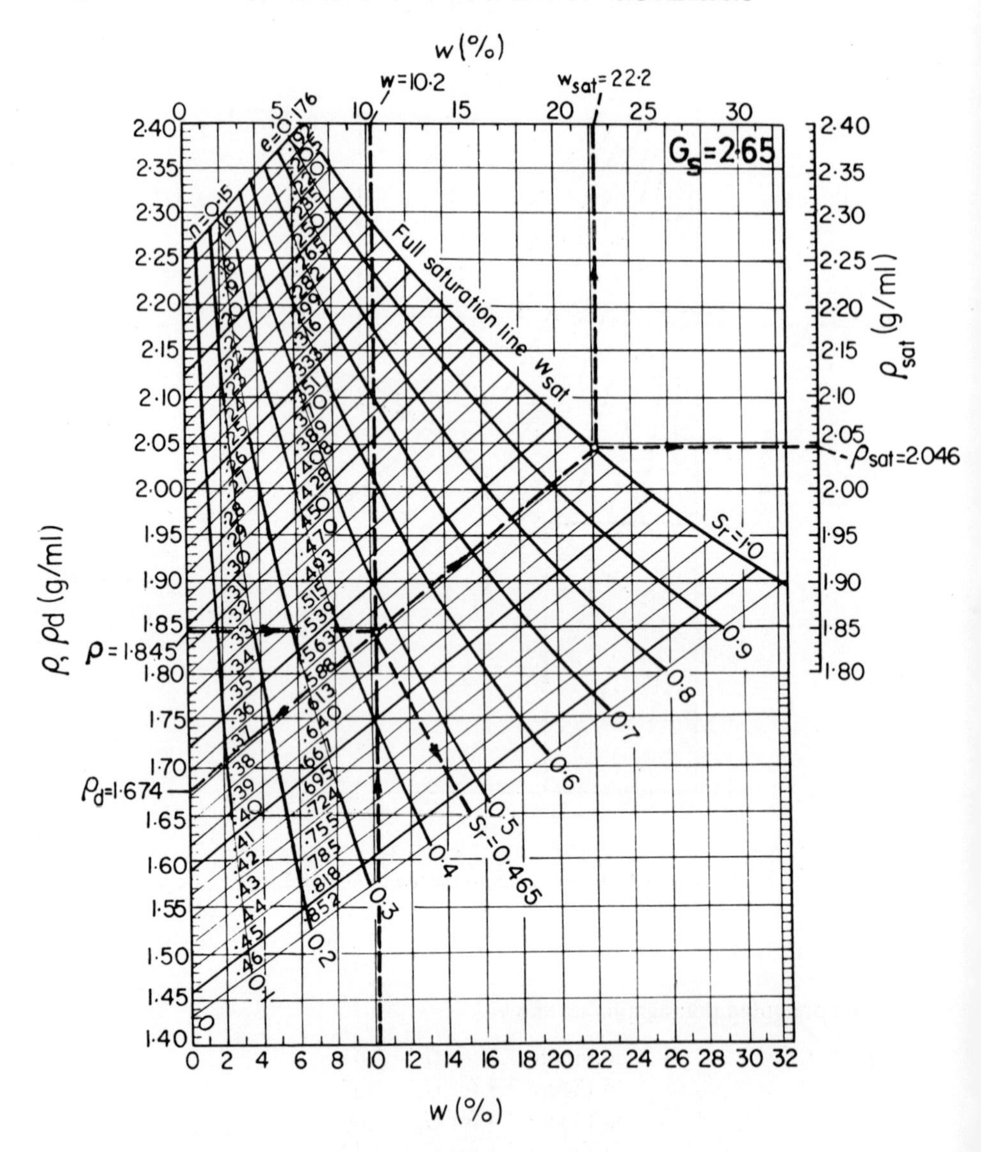

Figure 7.2. Nomogram for determination of n, e, ρ_d, ρ_{sat}, w_{sat}, and S_r from the knowledge of ρ and w and for $G_s = 2{\cdot}65$.

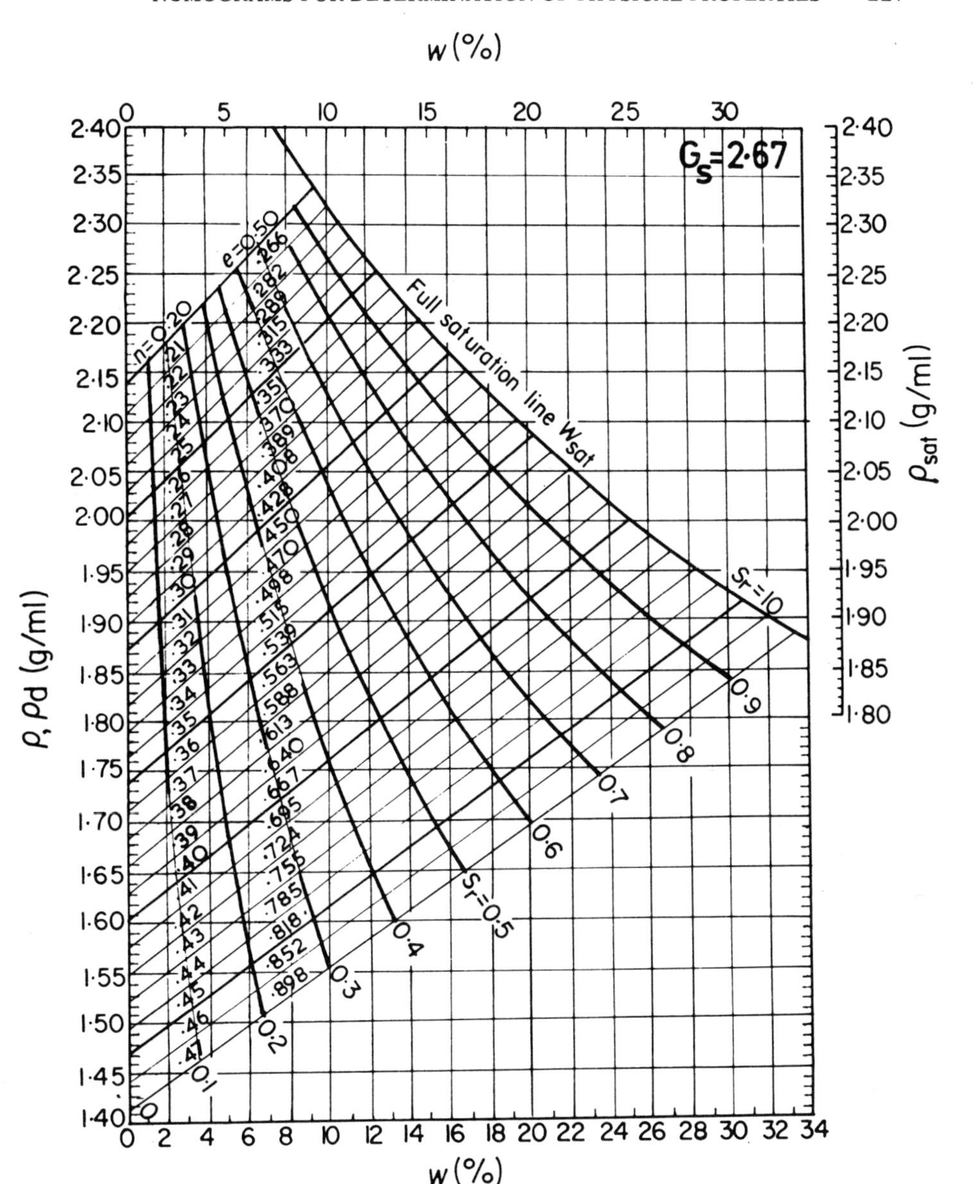

Figure 7.3. Nomogram for determination of n, e, ρ_d, ρ_{sat}, w_{sat}, and S_r from the knowledge of ρ and w and for G_s = 2·67.

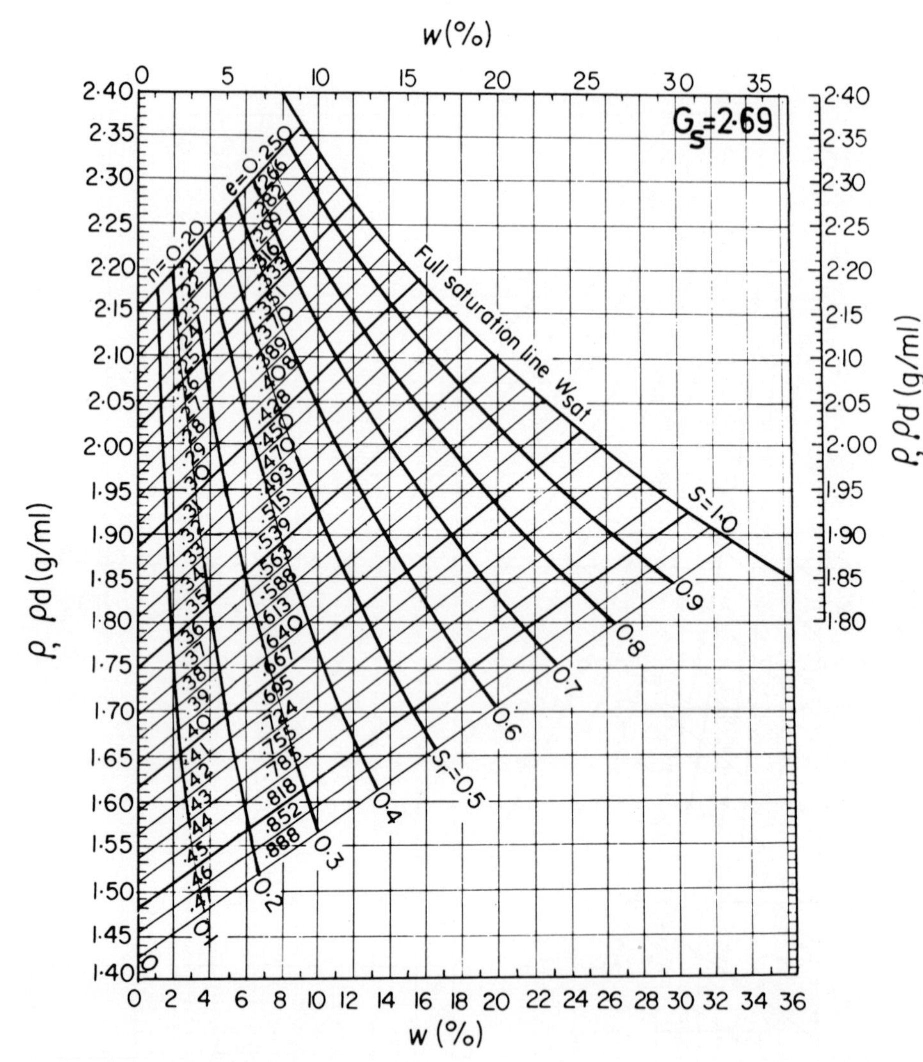

Figure 7.4. Nomogram for determination of n, e, ρ_d, ρ_{sat}, w_{sat}, and S_r from the knowledge of ρ and w and for $G_s = 2{\cdot}69$.

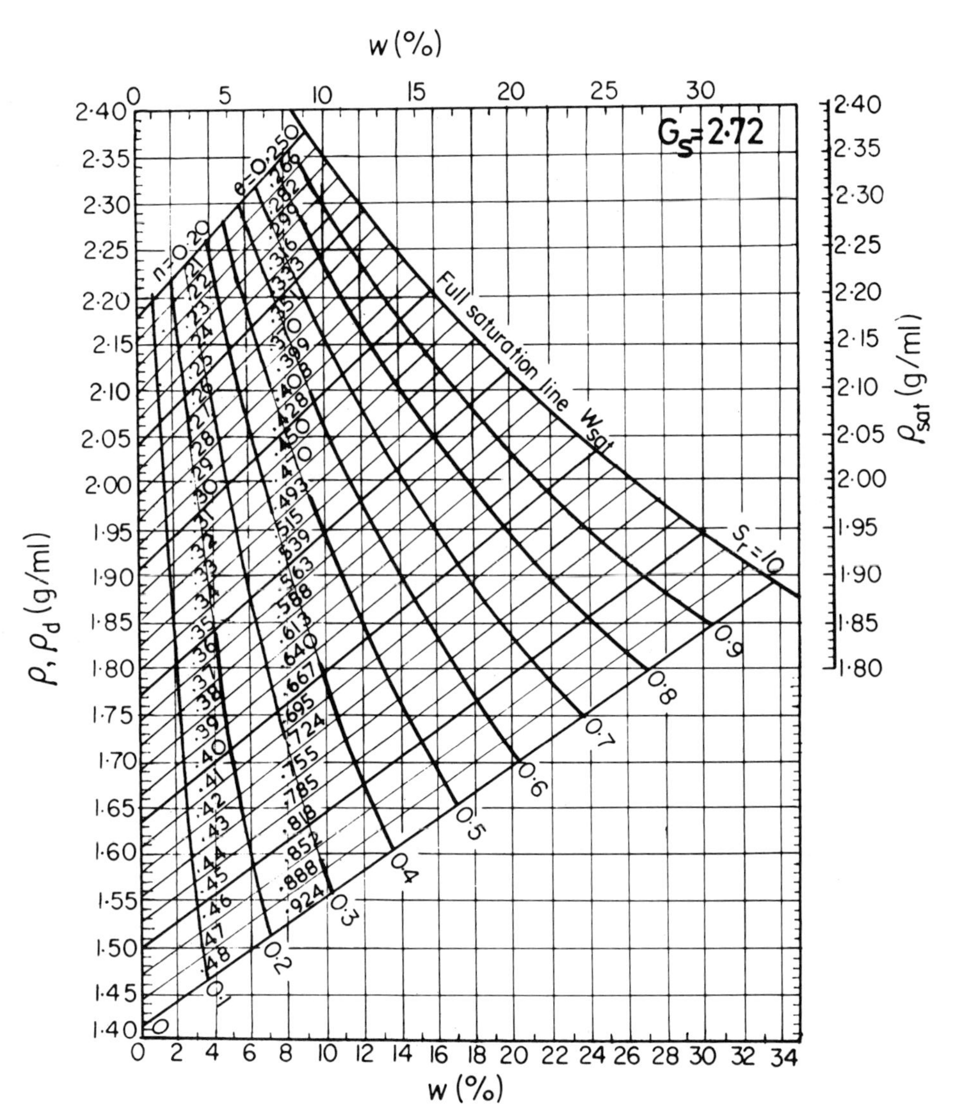

Figure 7.5. Nomogram for determination of n, e, ρ_d, ρ_{sat}, w_{sat}, and S_r from the knowledge of ρ and w and for $G_s = 2{\cdot}72$.

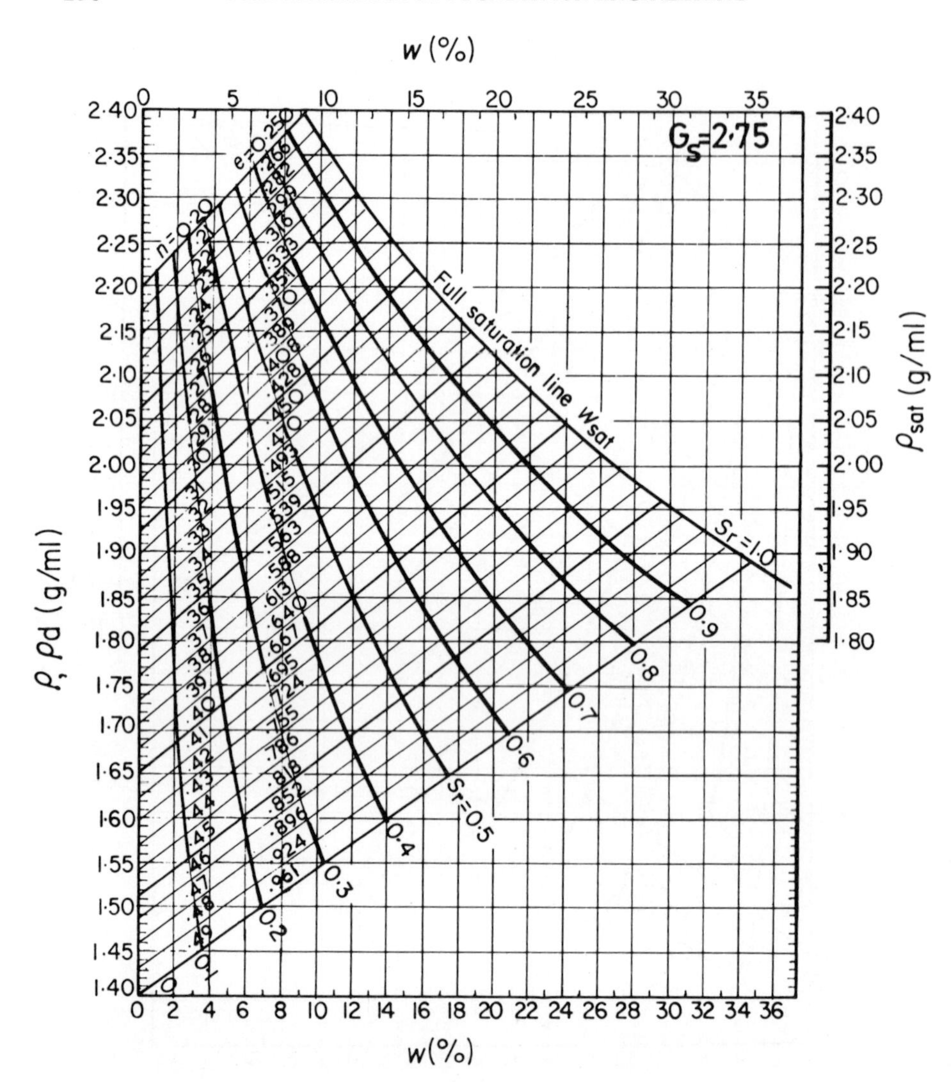

Figure 7.6. Nomogram for determination of n, e, ρ_d, ρ_{sat}, w_{sat}, and S_r from the knowledge of ρ and w and for G_s = 2·75.

Volume 1: References

Akroyd, T. N. W., 1957, *Laboratory Testing in Soil Engineering*, Geotechnical Monograph No. 1, Soil Mechanics Ltd., London.

Babkov, V. F., and Gerburt-Gejbovich, A. V., 1956, *Osnovy Gruntovedenya Mekhaniki Gruntov*, Dorizdat, Moscow.

Bernatzik, W., 1947, *Baugrund und Physik*, Zürich.

Bishop, A. W., and Henkel, D. J., 1957, *The Measurement of Soil Properties in Triaxial Test*, Edward Arnold, London.

Bjerrum, L., 1954, *Theoretical and Experimental Investigations on the Shear Strength of Soils*, Akademisk Trykningssentral, Oslo.

——1967, 'Progressive failure in slopes of overconsolidated plastic clays and clay shales', *Proc. Am. Soc. Civil Engrs. (J. Soil Mech. Found. Div.)*, **93**, No. SM5, Part 1, 1–49.

Borowicka, H., 1961, 'Über die Scherfestigkeit bindiger Böden', *Mitt. Inst. Grundbau und Bodenmech., Vienna*, **3**.

——1963, 'Vienna method of shear testing', *Proc. Symp. Lab. Shear Testing of Soils, Ottawa, A.S.T.M. Spec. Tech. Publ.*, No. 361, 306–14.

British Standard 1377, 1967, *Methods of Testing Soils for Civil Engineering Purposes*, British Standards Institution, London.

British Standard Code of Practice CP 2001, 1957, *Site Investigations*, The Council for Codes of Practice, British Standards Institution, London.

British Standard Code of Practice CP 2004, 1971 Draft, *Foundations*, The Council for Codes of Practice, British Standards Institution, London.

Casagrande, A., 1936, 'The determination of the preconsolidation load and its practical significance', *Proc. 1st Intern. Conf. Soil Mech. Found Eng., Cambridge, Mass.*, Vol. 3, pp. 60–4.

——1948, 'Classification and identification of soils', *Trans. Am. Soc. Civil Engrs.*, **113**, 901–30.

Casagrande, L., 1949, 'Electro-osmosis in soils', *Geotechnique*, **3**, 150–77.

Croney, D., and Coleman, J. D., 1961, 'Pore pressure and suction in soils', *Proc. Pore Pressure and Suction Conf., London*, pp. 31–7.

Cytovich, N. A., 1968, *Mekhanika Gruntov*, Izdatelstvo 'Vysshaya Shkola', Moscow.

Davis, E. H., and Poulos, H. G., 1968, 'The use of elastic theory for settlement prediction under three-dimensional conditions', *Geotechnique*, **18**, 67–91.

Dumbleton, M. J., and West, G., 1971, 'Preliminary sources of information for site investigation in Britain', *Road Res. Lab. Rept., Crowthorne*, No. RL 403.

Endell, K., Breth, H., and Loes, W., 1939, *Zusammenhang zwischen Kolloid-chemischen sowie bodenphysikalischen Kennzifern bindiger Bödern und Frostwirkung*, Berlin.

Gibbs, H. J., and Holtz, W. G., 1957, 'Research on determining the density of sands by spoon penetration testing', *Proc. 4th Intern. Conf. Soil Mech. Found. Eng., London*, Vol. 1, p. 35.

Gibson, R. E., and Anderson, W. F., 1961, '*In situ* measurement of soil properties with pressuremeter', *Civil Eng. Publ. Works Rev., London*, May, 615–18.

Glossop, R., and Skempton, A. W., 1945, 'Particle size in silts and sands', *J. Inst. Civil Engrs. (London)*, Paper 5492, Dec., 81–105.

Golder, H. Q., and Gass, A. A., 1962, 'Field test for determination of permeability of soil strata', *Field Testing of Soils, ASTM Spec. Tech. Publ.*, No. 322, 29–46.

Hogentogler, C. A., 1937, *Engineering Properties of Soil*, McGraw-Hill, New York.

Horn, A., 1964, *Die Scherfestigkeit von Schluff,* Westdentschner Verlag, Köln.

Hvorslev, M. J., 1936, 'Über die Festigkeitseigenschaften Gestörter Bindiger Böden', *Ingeniorvidenskab. Skrifter, Ser. A*, No. 45.

——1949a, *Subsurface Exploration and Sampling of Soils for Civil Engineering Purposes*, U.S. Army Waterways Experimental Station Corps of Engineers, Vicksburg, Miss.

——1949b, *Time Lag in the Observation of Ground Water Levels and Pressures*, U.S. Army Waterways Experimental Station Corps of Engineers, Vicksburg, Miss.

Keil, K., 1954, *Ingenieurogeologie und Geotechnik,* Knapp Verlag, Halle.

Kérisel, J., and Quatre, M., 1968, 'Settlement under foundations', *Civil Eng. Public Works Rev., London,* May and June, 531–5, 661–6.

Kézdi, A., 1962, *Erddrucktheorien*, Springer-Verlag, Berlin.

——1964, *Bodenmechanik*, Vol. 1, Verlag f. Bauwesen, Berlin, p. 150.

Kirsch, H., 1968, *Applied Mineralogy*, Chapman and Hall, London.

Kollbrunner, C. F., 1956, *Fundation und Konsolidation*, SDV-Fachbücher, Zürich.

Komornik, A., Wiseman, G., and Frydman, S., 1969, 'A study of *in situ* testing with the pressuremeter', *Conf. in situ investigations in soils and rocks, Brit. Geotech. Soc., London*, pp. 97–106.

Lambé, T. W., 1951, *Soil Testing for Engineers*, Wiley, New York.

——1953, 'The structure of inorganic soil', *Proc. Am. Soc. Civil Engrs.*, **79**, No. 315, Oct.

——1967, 'The stress path method', *Proc. Am. Soc. Civil Engrs. (J. Soil Mech. Found. Div.)*, **93**, No. SM6, 309–31.

Lambé, T. W., and Whitman, R. V., 1969, *Soil Mechanics*, Wiley, New York.

Laudon, A. G., 1952, 'The computation of permeability from simple soil tests', *Geotechnique*, **3**, 165–83.

Litvinov, J. M., 1951, *Issledovanya Gruntov v Polovykh Usloviyakh*, Kharkov, Moscow.

Mackey, R. D., 1964, 'The shearing resistance of granular soils', *Civil Eng. Public Works Rev., London,* Sep.

Palmer, D. J., and Stuart, J. G., 1957, 'Some observations on the standard penetration test and a correlation of the test with a new penetrometer', *Proc. 4th Intern. Conf. Soil Mech. Found. Eng., London,* Vol. 1, pp. 231–6.

Piaskowski, A., 1954, 'Badania nad elektroosmotycznym przepływem wody w gruntach', *I.T.B. Publ. Ser. B, Budownictwo i Architektura, Warsaw,* No. 6. 1956, 'Badania nad tiksotropia zawiesin iłowych i ich zastosowaniem w budownictwie', *I.T.B. Publ. Ser. I, Warsaw,* No. 3.

Polish Standard PN-54/B-02480, 1954, *Grunty Budowlane Klasyfikaeja (Engineering Soils, Classification)*, Wydawnictwa Normalizacyjne, Warsaw.

Polish Standard PN-55/B-04481 to 04494, 1955, *Grunty Budowlane, Klasyfikacja Właściwości Fizycznych (Engineering Soils, Determination of Physical Properties)*, Wydawnictwa Normalizacyjne, Warsaw,

Polish Standard PN-59/B-03020, 1959, *Grunty Budowlane, Wytyczne Wyznaczania Dopuszczalnych Obciazen Tednosthowych (Engineering Soils, Directives for Determination of Allowable Bearing Stresses) Wydawnictwa Normalizacyjne.*

Preece, E. F., 1947, 'Geotechnics and geotechnical research', *Proc. 27th Ann. Meeting Highway Res. Board, Wash., D.C.,* pp. 384–416.

Priklonskii, V. A., 1949, *Gruntovedenye*, G.J.G.L., Moscow.

Quincke, G., 1861, 'Über die Fortführung materieller Tielchen durch strömende Elektrizität', *Poggendorffs Ann. Physik Chemie*, **113**, No. 8.

Reuss, F. F., 1809, 'Sur un nouvel effet de l'électricité galvanique', *Mem. Soc. Imp.,* **1**, 327.

Road Research Laboratory, D.S.I.R., 1955, *Soil Mechanics for Road Engineers,* H.M.S.O., London.

Rode, A. A., 1955, *Vodnyye Svoystva Pochv i Gruntov*, Akademiya Nauk U.S.S.R.

Rowe, P. W., 1962, 'The stress–dilatancy relation for static equilibrium of an assembly of particles in contact', *Proc. Roy. Soc. (London), Ser. A*, **269**, 500–27.

Rowe, P. W., and Barden, L., 1966, 'A new consolidation cell', *Geotechnique*, **16**, 162–70.

Roza, S. A., 1950, 'Osadki gidrotekhnicheskikh sooruzhenii na glinakh s maloi vlazhnostyu', *Gidrotekh. Stroit.*, No. 9.

———1962, *Mekhanika Gruntov*, Vysshaya Shkola, Moscow.

Sanglerat, G., 1965, *Le Penetrometre et la Reconnaissance des Sols*, Dunod, Paris.

Simons, N. E., and Som, N. N., 1969, 'The influence of lateral stresses on the stress deformation characteristics of London Clay', *Proc. 7th Intern. Conf. Soil Mech. Found. Eng., Mexico*, Vol. 1, pp. 369–77.

Skempton, A. W., 1948, 'Vane tests in the alluvial plain of the river Forth near Grangemouth', *Geotechnique,* **1**, 111–24.

———1953, 'The colloidal "activity" of clays', *Proc. 3rd Intern. Conf. Soil Mech. Found Eng., Zürich*, Vol. 1, pp. 57–61.

———1954, 'The pore pressure coefficients *A* and *B*', *Geotechnique*, **4**, 143–7.

———1960, 'Horizontal stresses in clay strata', *Proc. Midland Soil Mech. Found.*

———*Eng. Soc., England*, Vol. 3, No. 20.

1964, 'Long-term stability of clay slopes', *Geotechnique*, **14**, 77–102.

Skempton, A. W., and Bishop, A. W., 1950, 'The measurement of the shear strength of soils', *Geotechnique*, **2**, 90–107.

Skempton, A. W., and Bjerrum, L., 1957, 'A contribution to the settlement analysis of foundations on clay', *Geotechnique*, **7**, 168–78.

Skempton, A. W., and Hutchinson, J., 1969, 'Stability of natural slopes and embankment foundations', *Proc. 7th Intern. Soil Mech. Found. Eng., Mexico*, State of the Art' Vol., pp. 291–340.

Skempton, A. W., and Sowa, V., 1963, 'The behaviour of saturated clays during sampling and testing', *Geotechnique*, **3**, 269–90.

Strzelecki, A., and Dudek, J., 1970, 'The use of pressure meter for *in situ* soil testing (in Polish)', *Inzynieria i Budownictwo*, April, 141–6.

Taylor, D. W., 1948, *Fundamentals of Soil Mechanics*, Wiley, New York.

Terzaghi, K., 1943, *Theoretical Soil Mechanics*, Wiley, New York.

Terzaghi, K., and Peck, G. B., 1967, *Soil Mechanics in Engineering Practice*, 2nd edn., Wiley, New York.

Troickaya, M. N., 1961, *Posobiye k Laboratornym Rabotam po Mekhanike Gruntov*, Izdatelstvo Moskovskogo Universiteta, Moscow.

Trollope, D. H., 1960, 'The fabric of clays in relation to shear strength, *Proc. 3rd Australia–New Zealand Conf. Soil Mech. Found Eng.*, pp. 197–202.

Ward, W. H., Burland, J. B., and Gallois, R. W., 1969, 'Geotechnical assessment of a site at Mundford, Norfolk, for a large proton accelerator', *Bld. Res. Sta., Current Paper,* No. 3/69.

Ward, W. H., Burland, J. B., and Gallois, R. W., 1969, 'Geotechnical assessment properties of London Clay', *Geotechnique*, **9**, 33–58.

WGT Gidroproject, 1950, 'Inzhenerno–geologicheskiye issledovanya dlya gidroenergotcheskogo stroitelstva, *Gosgeolizdat.*

Wiłun, Z., 1947, *Gruntoznawstwo Drogowe*, I.T.B., Warsaw.

———1951, 'Makroskopowa metoda oznaczania stanu konsystencji gruntów spoistyck', *Inzynieria i Budownictwo*, July, Aug.

———1955a, *Tymczasowe Wytyczne Laboratoryjnego Oznaczania Modułow Ścisliwości Gruntów*, I.T.B., Warsaw.

———1969, *Mechanika Gruntów i Gruntoznowstwo Drogowe,* 3rd edn., W.K.L., Warsaw.

Winterkorn, H. F., 1955, 'The science of soil stabilization', *Highways Res. Board Bull., Natl. Acad. Sci., Wash., D.C.,* No. 108.

Zielinski, K., 1956, *Zeskalanie Gruntów Metoda Cerbertowicza*, P.W.N., Warsaw.

Zinkin, G. N., 1956, 'Nekotoryye itogi proizvodstwennogo primeneniya elektrokhimicheskogo zakreplenya gruntov', *Sb. Tr., Leningr. Inst. Inzh. Zheleznodor. Transp.*, No. 150.

Relationship Between S.I., Metric, and Imperial Units of Measurement

The *basic units* of the S.I. system (Système International d'Unités) commonly used in Civil Engineering are as follows:

(i) Length: metre (m)
(ii) Mass: kilogramme (kg)
(iii) Time: second (s)
(iv) Temperature: degree Kelvin (°K) equal to the commonly used degree Celcius (°C).

Unit of	S.I. units		Metric units	Imperial units (1 foot = $\frac{1}{3}$ Imp. yard)
	Basic	Recommended sub- and multiples		
Length	10^{-9}m	1 nm (nanometre)	1 mμ (millimicron)	metric units usually used
	10^{-6}m	1 μm (micrometre)	1 μ (micron)	
	10^{-3}m	**1 mm (millimetre)***	1 mm	0·03937 in (inch)
	10^{-2}m	not recommended	1 cm (centimetre)	0·3937 in
	1 m	**1 m (metre)†**	1 m	39·37 in; 3·281 ft (feet)
	10^{3}m	1 km (kilometre)	1 km	3281 ft
Area	10^{-6}m^2	**1 mm^2***	1 mm^2	1·550 x 10^{-3} in^2
	10^{-4}m^2	not recommended	1 cm^2	0·1550 in^2
	1 m^2	**1 m^2†**	1 m^2	10·764 ft^2
	10^{4}m^2	1 ha (hectare)	1 ha	2·471 acres
Volume or capacity	10^{-6}m^3	**1 ml*** (millilitre)	1 cm^3	0·061 in^3
	10^{-3}m^3	1 l (litre)	10^3cm^3 or 1 l	61·025 in^3; 0·220 gal 0·2642 U.S. gal
	1 m^3	**1 m^3†** or 10^3l	1 m^3 or 10^3l	35·315 ft^3

Unit of	S.I. units		Metric units	Imperial units (1 foot = $\frac{1}{3}$ Imp. yard)
	Basic	Recommended sub- and multiples		
Velocity, rate	10^{-3} m/s 10^{-2} m/s 1 m/s	**1 mm/s**‡ not recommended **1 m/s**†	10^{-1} cm/sec 1 cm/sec 1 m/sec	0·03937 in/sec 0·03281 ft/sec 3·281 ft/sec
Rate of flow	10^{-6} m^3/s 10^{-3} m^3/s 1 m^3/s	**1 ml/s*** **1 l/s**‡ **1 m^3/s**† or 10^3 l/s	1 cm^3/sec 1 l/sec 1 m^3/sec or 10^3 l/sec	3·531 x 10^{-3} ft^3/sec 0·03531 ft^3/sec 35·315 ft^3/sec
Mass	10^{-3} kg 1 kg 10^3 kg	**1 g*** (gramme) **1 kg**‡ **1 t**† (tonne)	1 gm or gr 1 kg 1 t	2·205 x 10^{-3} lb (pound) 2·205 lb 2205 lb, 0·9842 T (tons) 1·1023 short tons
Density	1 kg/m^3 10^3 kg/m^3	**1 kg/m^3**† 1 kg/l or **1 g/ml*** or **1 t/m^3**†	1 kg/m^3 1 kg/l or 1 gm/cm^3 or 1 t/m^3	0·06243 lb/ft^3 62·4 lb/ft^3
Force	1 kgm/s^2 10^3 kgm/s^2 10^6 kgm/s^2	**1 N*** (newton) **1 kN**† (kilonewton) 1 MN (Meganewton)	0·101971 kgf 101·971 kgf 101·971 tf	0·22481 lbf 224·81 lbf 117·40 Tf
Stress, pressure	1 kg/ms^2 10^3 kg/ms^2 10^6 kg/ms^2	1 N/m^2 **1 kN/m^2** ‡ **1 MN/m^2**‡ or 1 N/mm^2	1·01971 x 10^{-5} kgf/cm^2 101·971 kgf/cm^2 101·971 tf/cm^2	0·02088 lbf/ft^2 0·14504 lbf/in^2 20·885 lbf/ft^2 9·3238 Tf/ft^2
Unit weight (specific weight)	1 kg/m^2s^2 10^3 kg/m^2s^2 10^6 kg/m^2s^2	1 N/m^3 **1 kN/m^3** ‡ 1 MN/m^3	1·0197 x 10^{-4} gf/cm^3 0·10197 gf/cm^3 101·97 kgf/m^3 101·97 tf/m^3	6·3657 x 10^{-3} lbf/ft^3 6·3657 lbf/ft^3 2·8418 Tf/ft^3
Temperature	1°K (Kelvin)	1°C (Celsius)	1°C	9/5°F

* Recommended for laboratory work.
† Recommended for design work.
‡ Recommended for both above uses.

Gravitational acceleration (standard average) $g = 9{\cdot}80665\ m/s^2 = 32{\cdot}174\ ft/sec^2$

Approximate Conversion Factors

1. Length: 1 mm ≈ 1/25 in; 1 in ≈ 25 mm; 1 m ≈ 10/3 ft; 1 ft ≈ 3/10 m.
2. Volume: 1 gal ≈ 4·5 l.
3. Mass: 1 kg ≈ 2·2 lb; 1 t ≈ 1 T (Imperial).
4. Density: 1000 kg/m^3 = 1 g/ml = 1 t/m^3 = 62·4 lb/ft^3; 125 lb/ft^3 ≈ 2000 kg/m^3.
5. Force: 1 N ≈ 1/10 kgf ≈ 2/9 lbf; 1 MN ≈ 100 Tf (Imperial) ≈ 100 tf; 1 Tf ≈ 10 kN.
6. Stress, pressure: 1 lbf/in^2 ≈ 7 kN/m^2; 1 Tf/ft^2 ≈ 100 kN/m^2; 1 lbf/ft^2 ≈ 1/20 kN/m^2.
7. Unit weight: 1 kN/m^3 ≈ 100 kgf/m^3 ≈ 6·5 lbf/ft^3.

S.I. Units in Foundation Engineering

1. Linear dimensions.
   (a) Laboratory work, details of apparatus, etc. – mm.
   (b) Foundation and earthwork details – m.
2. Density.
   (a) Laboratory calculations – g/ml.
   (b) Foundation and earthwork calculations – t/m^3 or kg/m^3.
3. Unit weights – kN/m^3.
4. Seepage.
   (a) Coefficient of permeability – mm/s or m/s.
   (b) Rate of flow – ml/s, l/s, or m^3/s.
5. Stress and pressure.
   (a) Normal and shear stresses in all calculations – kN/m^2.
   (b) Deformation and stiffness moduli and modulus of elasticity – MN/m^2.
6. Consolidation.
   (a) Coefficient of volume compressibility – m^2/MN.
   (b) Coefficient of consolidation – mm^2/s or m^2/s.

Notes for American Readers

For American readers of this book the authors have included a list of ASTM Standards relevant to the material contained in the text.

It must be emphasised, however, that the book is based on sound engineering principles which are universal in character, and therefore, it is in no way restricted to any one country. In fact the book presents an unusual combination of the experiences of two authors who have studied, taught and practised the subject in countries in which its development proceeded on fairly independent lines.

Z. Wiłun and K. Starzewski

List of American Society for Testing and Materials (ASTM) Standards Relevant to the Material Contained in this Book

The first number (e.g. D420) indicates the fixed designation of the ASTM Standard or Tentative; the number immediately following the designation indicates the year of original adoption or, in the case of revision, the year of last revision. A number in parentheses indicates the year of last reapproval. Tentatives are identified by the letter T.

A. ASTM Standards Relevant to Classification of Soils and Determination of their Physical Properties

D 2487–69	Classification of Soils for Engineering Purposes.
D 2488–69	Rec. Practice for Description of Soils (Visual–Manual Procedure).
D 653–67	Def. of Terms and Symbols Relating to Soil and Rock Mechanics.
D 2607–69	Classification of Peats, Mosses, Humus, and Related Products.
D 421–58 (1965)	Dry Preparation of Soil Samples for Particle-size Analysis and Determination of Soil Constants.
D 422–63	Particle-size Analysis of Soils.
D 423–66	Test for Liquid Limit of Soils.
D 424–59 (1965)	Test for Plastic Limit and Plasticity Index of Soils.
D 427–61 (1967)	Test for Shrinkage Factors of Soils.
D 854–58 (1965)	Test for Specific Gravity of Soils.
D 1140–54 (1965)	Test for Amount of Material in Soils Finer than the No. 200 Sieve.
D 2049–69	Test for Relative Density of Cohesionless Soils.
D 2216–66	Laboratory Determination of Moisture Content of Soil.
D 2217–66	Wet Preparation of Soil Samples for Grain-size Analysis and Determination of Soil Constants.
D 2419–69	Test for Sand Equivalent Value of Soils and Fine Aggregate.
D 2434–68	Test for Permeability of Granular Soils (Constant Head).

B. ASTM Standards Relevant to Strength and Compressibility of Soils and Rock

D 2166–66	Test for Unconfined Compressive Strength of Cohesive Soils.
D 2435–70	Test for One-dimensional Consolidation Properties of Soils.
D 2664–67	Test for Triaxial Compressive Strength of Undrained Rock Core Specimens without Pore Pressure Measurements.
D 2850–70	Test for Unconsolidated, Undrained Strength of Cohesive Soils in Triaxial Compression.

D 2938–71	Test for Unconfined Compressive Strength of Rock Core Specimens.

C. ASTM Standards Relevant to Soil Investigation Work and *In Situ* Testing

D 420–69	Rec. Practice for Investigating and Sampling Soils and Rock for Engineering Purposes.
D 1452–65	Soil Investigation and Sampling by Auger Boring.
D 1586–67	Penetration Test and Split-Barrel Sampling of Soils.
D 1587–67	Thin-walled Tube Sampling of Soils.
D 2113–70	Diamond Core Drilling for Site Investigation.
D 2573–67T	Field Vane Shear Test in Cohesive Soil.
D 2944–71	Method of Sampling Peat Materials.
D 1194–57 (1966)	Test for Bearing Capacity of Soil for Static Load on Spread Footings.
D 2487–69	Classification of Soils for Engineering Purposes.
D 2488–69	Rec. Practice for Description of Soils (Visual–Manual Procedure).
D 653–67	Def. of Terms and Symbols Relating to Soil and Rock Mechanics.

D. ASTM Standards Relevant to Compaction and Control of Compaction of Soils

D. 698–70	Test for Moisture-Density Relations of Soils, Using 5·5 lb Rammer and 12 inch Drop.
D 1556–64 (1968)	Test for Density of Soil in Place by the Sand-Core Method.
D 1557–64 (1968)	Test for Moisture–Density Relations of Soils Using 10 lb Rammer and 18 inch Drop.
D 1558–63 (1969)	Test for Moisture–Penetration Resistance Relations of Fine-grained Soils.
D 1883–67	Test for Bearing Ratio of Laboratory-Compacted Soils.
D 2167–66	Test for Density of Soil in Place by the Rubber-Baloon Method.
D 2168–66	Calibration of Mechanical Laboratory Soil Compactors.
D 2922–71	Determining the Density of Soil and Soil-Aggregate in Place by Nuclear Methods (Shallow Depth).
D 2937–71	Test for Density of Soil in Place by the Drive-Cylinder Method.

Subject Index

Author Index

(Page number in brackets refer to general references mentioned only in the list of references.)